2

AF708999

INTRODUCTION TO PIKE NATIONAL FOREST

The Pike National Forest is one of eleven National Forests in Colorado and one of 154 within the National Forest System of the United States. The Pike National Forest and the San Isabel National Forest are headquartered at 1920 Valley Drive, Pueblo, Colorado 81008. The 1,106,604 acres of Pike National Forest falls within Clear Creek, Teller, Park, Jefferson, Douglas and El Paso counties.

The recreation resource on the Pike National Forest is widespread and varied. Marking the north end of the is Mount Evans at 14,264 ft. elevation surrounded by the 73,000 acre Mount Evans Wilderness. Mount Bierstadt, another of Colorado's "fourteeners" at 14,060 ft elevation is also located in the Mount Evans Wilderness. North and west of South Park in the western part of the Forest, are the high elevations of the Mosquito Range from Guanella Pass south-ward to the Buffalo Peaks country. Except for Geneva and Jefferson Creeks, the recreation resource in this large expanse of Park County remains primarily primitive and undeveloped, accessible mainly by four-wheel drive, horse or foot. Remains of early day mining railroads and stage routes are still easily discernible, further adding to the attraction of the are More "fourteeners are found west of Fairplay: Mount Lincoln 14,286, Mount Bross 14,172 ft., and Mount Democrat 14,148 ft.

Also located in this portion of the Forest is the Windy-Ridge Bristlecone Pine Scenic Area. This scenic area, located about four miles northwest of Alma was established in July 1967 to protect a unique grove of beautifully grotesque and deformed bristlecone pine trees.

East of South Park but still in Park County are the popular fishing and camping areas of Lost Park, Tarryall Creek, and Elevenmile Canyon. Most of the Pike National Forest in this area remains undeveloped allowing the visitor to enjoy his recreation experience in a near natural environment.

Tarryall Creek and Lost Park are jump-off points for the Lost Creek Wilderness where motorized travel is prohibited. This approximately 106,000 acre area provides a place within an hours drive of the metropolitan area of Colorado Springs where an individual can experience a wilderness setting. The Lost Creek Wilderness contains bold outcrops of Pikes Peak granite that have been carved into an infinite variety of forms and shapes by the forces of erosion. Great dome like forms, spires, turrets and arete crests characterize the landscape. Talus slopes of huge boulders are common. Stream courses are tortuous and have in places carved underground channels through the talus debris. In addition to its great scenic beauty, the area offers the geologist, professional or amateur, an outstanding opportunity to study the effects of erosion of granite in a semi-arid climate.

Manitou Park is a popular area on the Pike National Forest in Teller County. This picturesque area of ponderosa pines and grass serves mainly as a "base camp" for visitors to the Pikes Peak Region. Numerous Forest Service campgrounds and commercial campgrounds are used to capacity from the middle of June to Labor Day. Manitou Lake Picnic Ground in the area is heavily used year around for fishing and picnicking. The remainder of National Forest land in Pikes Peak rising to 14,110 ft. elevation is the most dominant and well known recreation attraction on the National Forest in El Paso County. The Pikes Peak Toll Road which was completed in 1915, is operated by the City of Colorado Springs under permit from the Forest Service. Over 300,000 people use the toll road each year. Revenues from the tolls are used entirely for maintenance and improvement of the road and summit house. The famous "Cog Road" carries thousands to the top of the Peak. Thousands more hike up the Barr Trail from Manitou Springs. This 11.7 mile trail was designated a National Recreation Trail in 1979.

The Gold Camp and Rampart Range Roads are popular recreation roads just west of Colorado Springs on the Front Range. They provide opportunities for camping, picnicking and sightseeing as well as circle trips connecting Colorado Springs, Cripple Creek and Woodland Park.

The Pike National Forest offers sightseeing, hiking, lake fishing, boating, picnicking and cross-country skiing at the 500 acre Rampart Reservoir, a 60 minute drive from Colorado Springs. Also, at the area is the Business and Professional Women's Club Nature Trail designed for wheelchairs and with signing in Braille.

The Devil's Head Lookout Tower in Douglas County, the last operational lookout on the Front Range, sits on the highest point on the Rampart Range. This tower is accessible by the 1.3 mile long foot Devil's Head National Recreation Trail.

U.S. Forest Service

Any of the following offices will be happy to furnish further information about the ***PIKE NATIONAL FOREST.***

FOREST HEADQUARTERS **719 545-8737**
1920 Valley Drive Pueblo, CO 81008

SOUTH PARK RANGER DISTRICT **719 836-2031**
P.O. Box 219 Fairplay, CO 80440

PIKES PEAK RANGER DISTRICT **719 636-1602**
601 South Weber P.O. Box 25127
Colorado Springs, CO 80903

SOUTH PLATTE RANGER DISTRICT **303 275-5610**
Lakewood, CO 80225

TABLE of CONTENTS

How To Use This Guide

Trails described in the text are shown on the map page opposite the text. The map number shown on this page refers to the map number shown on the index map. The index map located in the front of the guide shows map numbers relative to the major roads and towns within the National Forest.

ACKNOWLEDGMENTS

WE WOULD LIKE TO THANK THE U. S. FOREST SERVICE FOR THEIR COOPERATION AND ASSISTANCE FOR MAKING THIS GUIDE POSSIBLE.

Most information contained in this document was compiled from Forest Service Recreation Opportunity Guides.

U.S. Geologic Survey
Colorado Division of State Parks and Outdoor Recreation

Cover Photo: **Tom Myers**
Cover Inset Photos: **PhotoDisc**
Cover Design: **Ken Grasman** - *Denver*
Map Graphics and Page Layout: **Michael Moore** - *Denver*

Editor & Publisher
Jack O. Olofson

Staff
Kristin Alexander
Ryan Alexander
Cynthia Alexander
Linnea Roberts
Dody Olofson

OUTDOOR BOOKS & MAPS

P.O. Box 417 Denver, Colorado 80201
Phone (303) 629-6111 • Fax (303) 628-9378

EXPLANATION

Map Scale: 1" = approximately 2 miles

Parking Area	Forest Service Facility	U.S. Highway	National Forest Area
Picnic Area	Fishing Area	State Highway	Water
Trail Head	Snowmobile Trail	County Highway	Trail
Downhill Ski Area	Ice Skating Area	Trail Number	River or Stream
Boat Launch	Cross Country Ski Area	Forest Service Road	Primary Road - Paved
Bicycle Trail	Hunting	Mountain	Improved Road - Unpaved
4WD Road	Campground	Colorado Trail	Unimproved Road - 4WD
Motorcycle Trail	Towns & Locales	Continental Divide Trail	Forest / Wilderness Boundary

Off-Highway Vehicles (OHV's)

On National Forest land all OHV's must be equipped with operating brakes, a muffler and a Forest Service-approved spark arrester. OHVs must also have and use a headlight and taillight between sunset and sunrise. OHVs should also be operated in a safe, quiet manner without disturbing the land, vegetation or wildlife.

Mountain Bikes:

Roads and most trails, with the exception of those in Wildernesses, are open for mountain bike use. Some trails are unsuitable due to steep grades, rocky conditions or stream crossings.

In order to have a safe and enjoyable ride, please keep the following suggestions in mind:

1. Trail challenges vary, so know the terrain and your skill level.
2. Be courteous to others on the trail.
3. When on the road, obey traffic regulations.
4. Stay on the road or trail, avoid shortcuts, switchbacks and damage to vegetation.
5. Wear a helmet.
6. Keep your bike in good repair and carry proper tools and spare parts.

Campgrounds:

Reservations: Although many Forest Service campgrounds are available on a first-come, first-served basis, selected sites and group areas are on a reservation system through a private corporation. The local District Ranger's office can tell you which campgrounds are on the reservation system, but they cannot make reservations for you. For information and reservations call Mistix 1-8OO-283-CAMP

Fees: Most campgrounds are fee areas. They will be clearly posted as such. Please follow posted instructions carefully and promptly. Closures: In some high-use areas, camping is restricted to developed sites in order to avoid erosion, sanitation, and fire safety problems. Please be alert for posted regulations in these high-use areas.

Dispersed Camping: When camping or picnicking outside developed sites, please use minimum-impact techniques. Leave a campsite as nice or nicer than you found it.

Wilderness:

In order to preserve wilderness areas for others to enjoy please follow these "no trace" camping guidelines.

1. Travel in small groups, avoiding heavy use areas.
2. Pack out metal, glass and other unburnables.
3. Don't bathe, wash dishes or camp within 100 feet of streams.
4. Don't tie horses or pack animals to trees.
S. Set up camp where you'll create the least disturbance.
6. Backpack stoves are recommended.
7. Human waste should be buried, and be at least 100 feet from streams.

General:

All Forest Service roads and trails that are open to motorized vehicles are signed on the ground with white arrows. This signing should be used to confirm the motorized travel status of all Forest routes.

Except as noted, the Pike National Forest is closed to cross-country motorized use in order to prevent resource damage, such as that caused by the unplanned development of new roads, and to reduce the disturbance of wildlife.

NUMERICAL TRAILS LISTING

Trails are in numerical order in document.

**** Trails open to trailbikes and ATV's less than 40" wide***

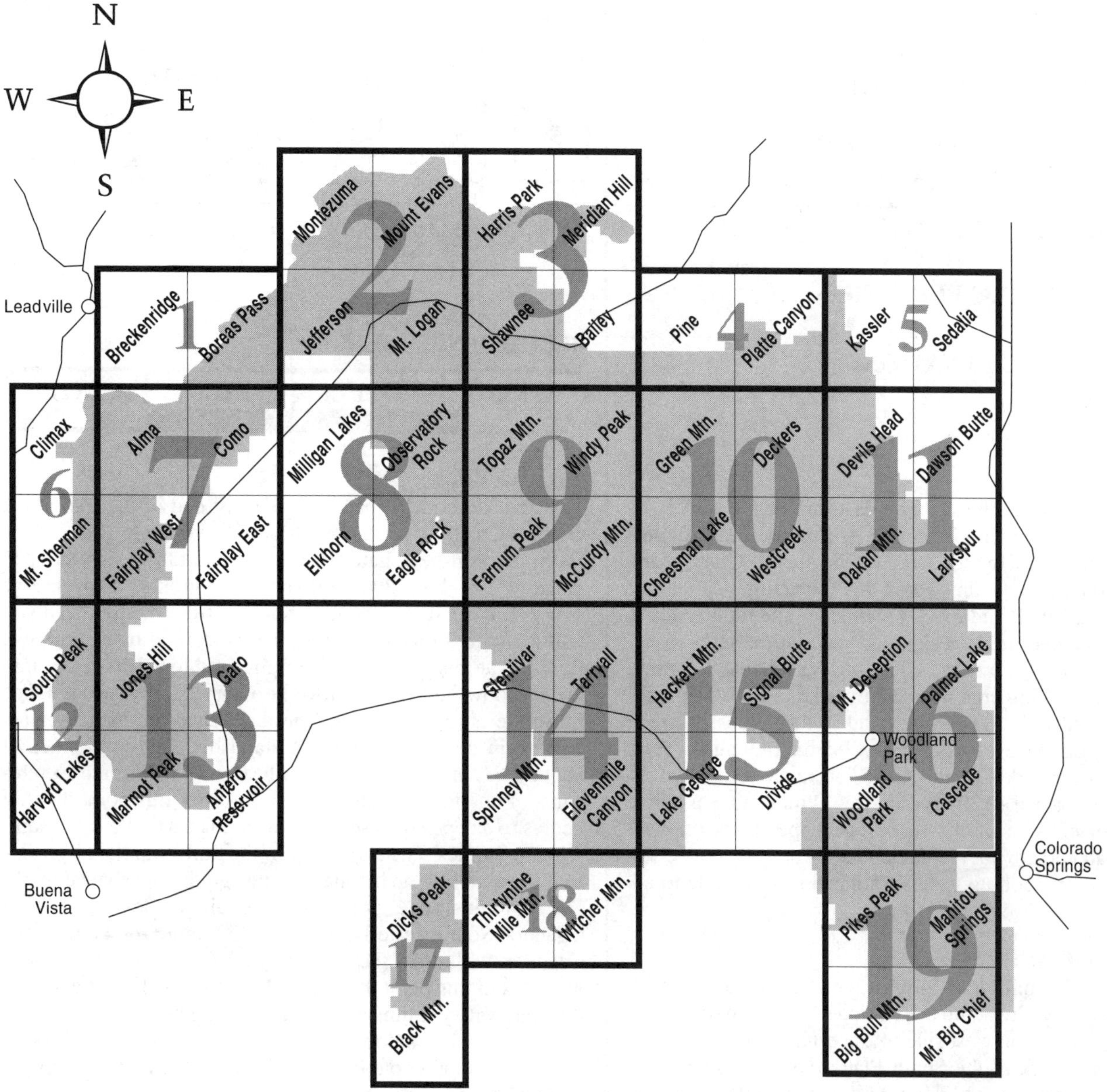

Map Numbers

Pike National Forest

(Showing USGS 7 1/2' quadrangle topographic map coverage)

Pike & San Isabel National Forest
Pikes Peak Ranger District
601 South Weber Street
Colorado Springs, Colorado 80903
(719) 636-1602

ROCKY MOUNTAIN REGION
MAP NUMBER(S): 2

NATIONAL FOREST: PIKE
RANGER DISTRICT: SOUTH PLATTE

USGS: MT. EVANS, MONTEZUMA, JEFFERSON QUADS

DIFFICULTY: EASY

DISTANCE: 5.5 MILES

TRAIL 601
BURNING BEAR

UPDATED: AUGUST 1990

USE: Moderate

TRAIL BEGINNING ELEVATION:
9,560 ft.

TRAIL ENDING ELEVATION:
9,545 ft.

RECOMMENDED SEASON:

SPRING	SUMMER	FALL	WINTER

ACCESS:
This trail has two trailheads. The east end of the trail is in Geneva Creek Valley and is located as follows. From Bailey drive west on U.S. Hwy 285 for 10.8 miles to Grant. Turn right (north) on Park County Road #62 (sometimes called the Guanella Pass Road) and drive 4.9 miles to the trailhead at a turn in the road. There is a sign and gate on the left (north) side of the road at the trailhead. Parking is available 0.2 miles further at the parking area for the Abyss Lake trail. The west end of the trail is in Hail Valley and is located as follows. From Bailey drive west on U.S. Hwy 285 for 14.9 miles. Turn right (north) on Park County Road #60 (also designated as Forest Service Road #120) and drive 3.0 miles to a parking area on the left side of the road. The upper parts of Park County Road #60 are not recommended for low clearance vehicles, but the portion up to this trailhead is accessible to all vehicles.

ATTRACTIONS:
This pleasant 5.5 mile trail follows the Burning Bear Creek west from Geneva Creek, up over a ridge, and down the Burning Bear Creek on the west side of the ridge as it drains west into the North Fork of the South Platte river in Hall Valley. The trail is accessible for hikers, horses and bicycles. Allow approximately 2 1/2 - 3 hours to hike the entire trail. The hike from the Geneva Creek Trailhead to the ridge summit will take about 90 minutes at a moderate pace.

NARRATIVE:
This trail description begins at the east end of the trail at the Geneva Creek Trailhead and proceeds west into Hall Valley. From the trail-head gate the trail leads northwest in an open meadow for about 200 yards and then crosses a big bridge over Geneva Creek. About 20 minutes from the trailhead you will come to a fork in the trail that is marked by a sign post. Bear left at this junction. (The right hand fork leads back into the meadow and eastward to the Guanella Pass Road.) The first mile is relatively flat. You then leave the meadow area and begin climbing through a stand of lodge-pole pine. At about 2 1/2 miles from the trailhead you will come to remnants of an old log cabin on the left Above this point the trail begins to steepen on a series of switchbacks to the ridge summit at 3.7 miles. The high point of the trail is marked by a six-foot dead tree stump having two metal diamond shaped markers. From this ridge (or saddle), the trail begins to descend in a general South westerly direction. 15 minutes down from the ridge you will cross the Burning Bear Creek as it now drains west. From this point the trail follows the creek all the way to the Hall Valley trailhead on what appears to be an old road. About 8 minutes below the stream crossing you will pass by remains of another old log :cabin. The trail is identified by diamond shaped metal markers nailed to trees along the way. Most of the markers are blue, the remainder being a gray color. The Hall Valley Trailhead (west end of the trail) marked by signs posted along Park County Road #60.

TRAIL PROFILE (ONE-WAY):

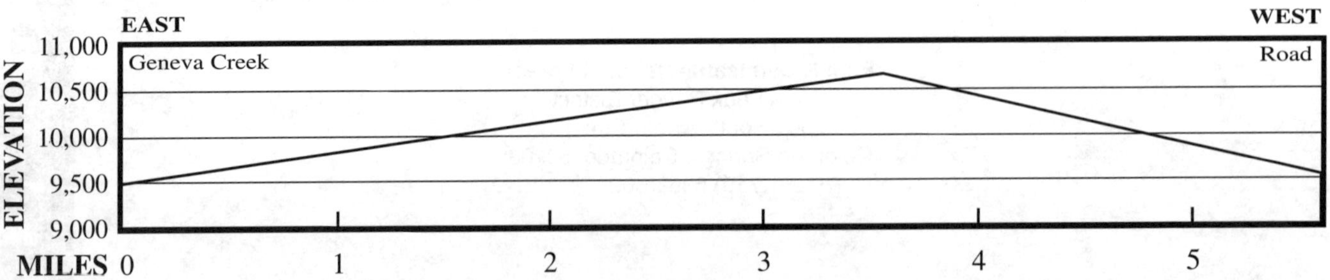

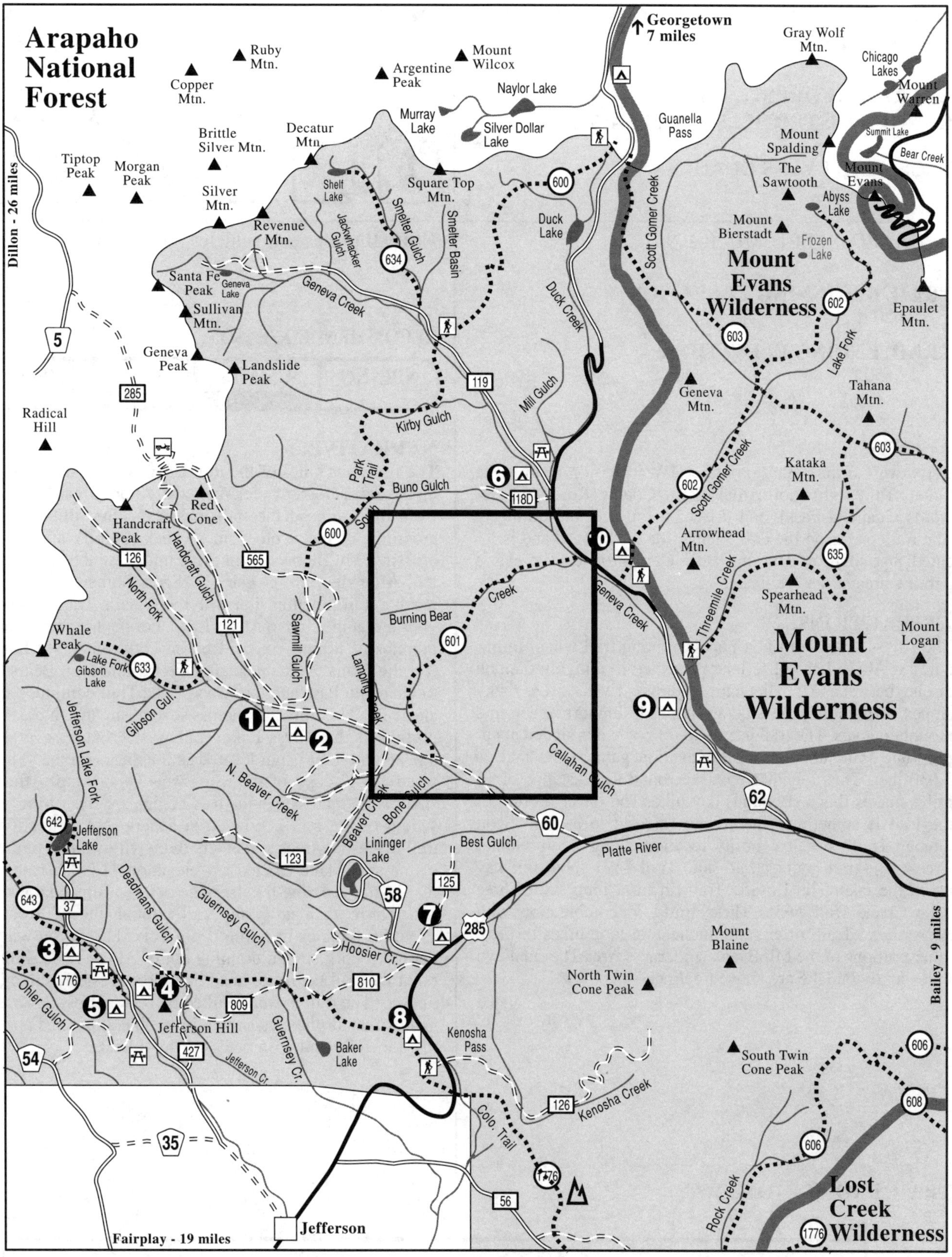

MAP 2

TRAIL #601

ROCKY MOUNTAIN REGION

MAP NUMBER(S): 2

NATIONAL FOREST: PIKE
RANGER DISTRICT: SOUTH PLATTE

USGS: MT. EVANS QUAD

DIFFICULTY: DIFFICULT

DISTANCE: 8.0 MILES

TRAIL 602
ABYSS LAKE

UPDATED: SEPTEMBER 1990

TRAIL BEGINNING ELEVATION:
9,620 ft.

TRAIL ENDING ELEVATION:
12,650 ft.

ACCESS:
Drive west from Bailey on U.S.. HWY 285 for 11 miles to Grant. Turn right (north) onto Park County Road #62 (towards Guanella Pass), and drive 5.5 miles to the trailhead. The trailhead is on the east (right) side of the road and is located just south of Burning Bear Campground. There is a parking area at the Trailhead.

ATTRACTIONS:
The Abyss Lake Trail is a popular hiking trail which terminates at Abyss Lake. The lake is nestled in a high mountain cirque between Mt. Bierstarit and Mt. Evans. The rocky slopes surrounding the lake are home to bighorn sheep and mountain goats. The trail is easy to follow and is signed at appropriate trail junctions. The elevation gain to the lake is 3,030 feet. This, together with the round trip distance of 16 miles makes this a strenuous day hike. The lower portion of the trail is sometimes incorrectly referred to as the "Scott Gomer Trail" because of its location along Scott Gomer Creek. The trail intersects Rosalie Trail #603 about midway up to the lake. The Rosalie Trail, in turn, joins with Three Mile Creek Trail #635. These trails, in combination with Abyss Lake Trail, offer some interesting loop hikes through other portions of the Mt. Evans wilderness Area The final two miles up to the lake are above treeline.

USE: Bicycles and vehicles prohibited.

RECOMMENDED SEASON:

SPRING	SUMMER	FALL	WINTER

NARRATIVE:
The first four miles of the trail heads in a northeast direction up the Scott Gomer Creek drainage. Approximately two miles from the trailhead the trail makes the first of three stream crossings of the Scott Gomer Creek. Shortly after this first crossing. Mt. Bierstadt will come into view ahead and on your left. After the third crossing of Scott Gomer Creek you will reach a trail junction sign with the Rosalie Trail. Continue to your left at the sign to Abyss Lake. For the next seven or eight minutes of hiking you will be on both the Abyss Lake and Rosalie Trails. At the next sign the Rosalie Trail Bears left up to Guanella Pass and the Abyss Lake Trail continues straight ahead in a N.E. direction across several streams. Note that this portion of the Abyss Lake Trail was rerouted several years ago and the trail is not located as it appears on the "Photo revised 1974" copy of the Mt. Evans Quad Map. After the 2 small stream crossings the trail begins a series of fairly steep switch backs, still bearing in a general N.E. direction. The trail leaves the trees and enters thick willows just to the S.W. of the small lake labeled as "elevation 11730" on the Quad map. After crossing the drainage outlet of this lake watch for rock cairns to identify the trail through the willows. After about 30 minutes of hiking from Lake 11730 you will cross the Lake Fork stream draining out of Abyss Lake. From this point the trail begins to head in a more westerly direction on tundra. The trail is well marked by rock cairns. Several remnants of an airplane crash will be visible along this portion of the trail. The trail terminates at Abyss Lake.

TRAIL PROFILE (ONE-WAY):

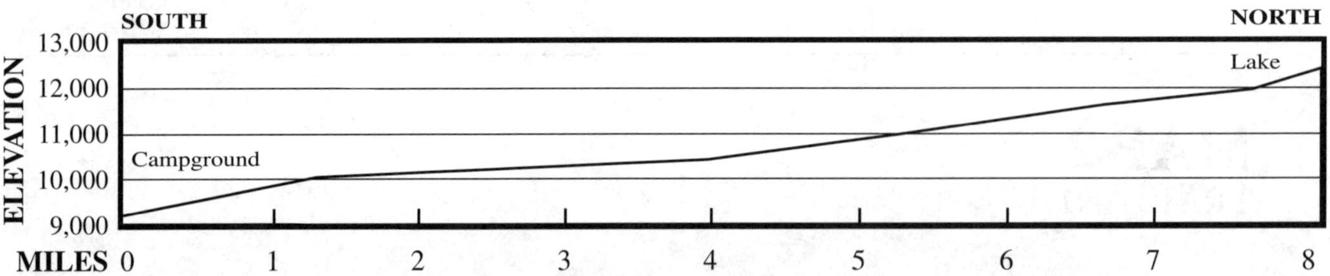

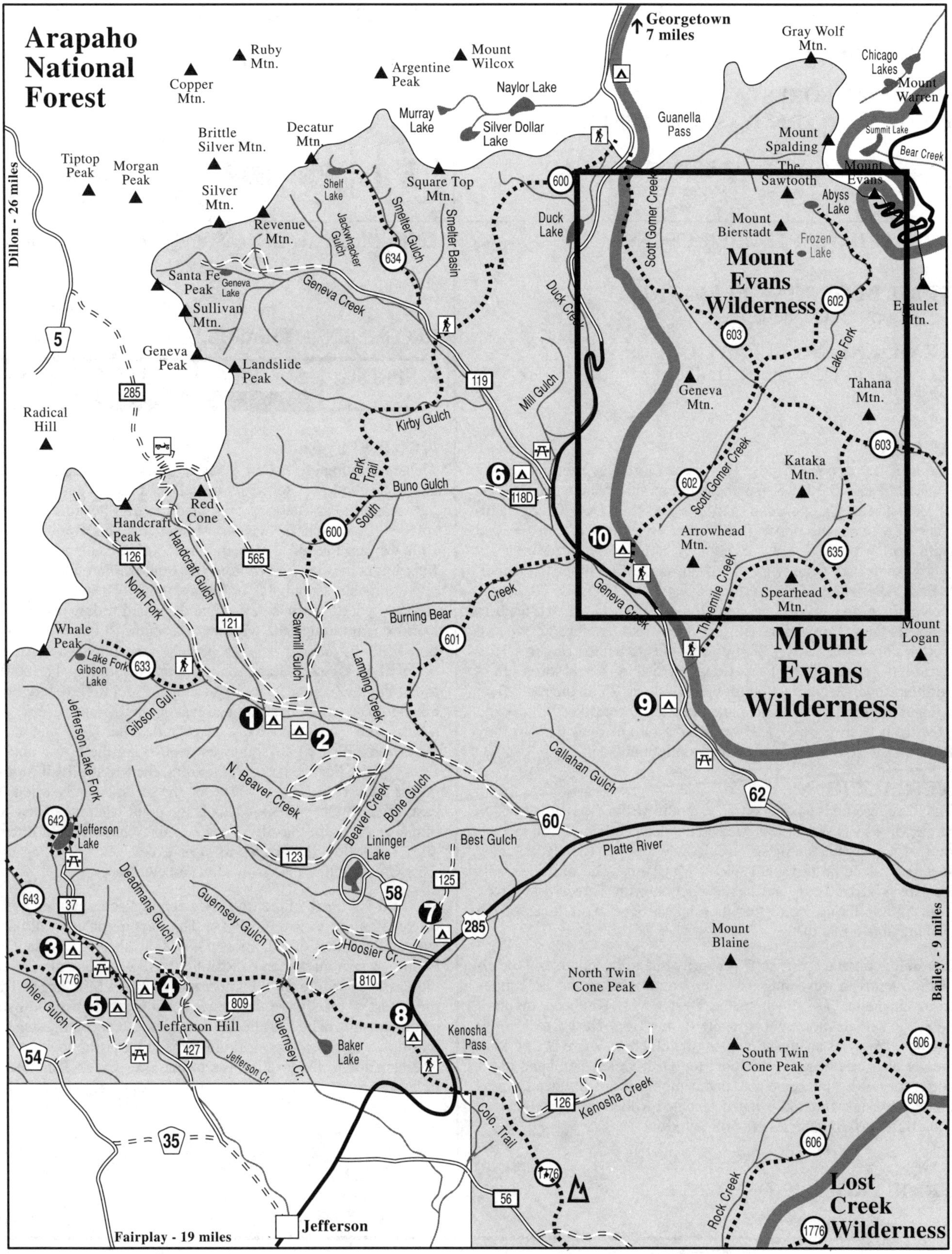

MAP 2

TRAIL #602

ROCKY MOUNTAIN REGION

MAP NUMBER(S): 2 and 3

NATIONAL FOREST: PIKE
RANGER DISTRICT: SOUTH PLATTE

USGS: MT. EVANS, HARRIS PARK QUADS

DIFFICULTY: DIFFICULT

DISTANCE: 12.2 MILES

TRAIL 603
ROSALIE

UPDATED: SEPTEMBER 1990

TRAIL BEGINNING ELEVATION:
11,669 ft. Guanella Pass

TRAIL ENDING ELEVATION:
9,320 ft. Deer Creek Trailhead

USE: Moderate, Bicycles and vehicles prohibited

RECOMMENDED SEASON:

SPRING	SUMMER	FALL	WINTER

ACCESS:

The trail has two trailheads. The upper (northwest) terminus is at Guanella Pass. The lower (southeast) terminus is about 1 mile west of Deer Creek Campground. and is called the "Deer Creek Trailhead". To reach the upper Trailhead at Guanella Pass, drive west from Bailey on U.S. Hwy 285 for 11 miles to Grant. Turn right (north) onto Park County Road #62 and drive 13.5 miles to Guanella Pass. Alternately, drive west from Denver on Interstate 70. Exit at Georgetown and follow the signs leading from Georgetown up to Guanella Pass. To reach the (lower) Deer Creek Trailhead drive east from Bailey on U.S. Hwy 285 for 2.5 miles to the top of Crow Hill. Turn left (north) onto Park County Road #43 and drive in a north/northwesterly direction for 6.8 miles to a "Y" in the road. Bear left and drive 2.1 miles to the parking area at the trailhead. If driving west from Denver on U.S. Hwy 285 the turnoff onto Park County Road #43 is 10.8 miles west of the stoplight in Conifer.

ATTRACTIONS:

The Rosalie Trail is a popular trail extending from the southeast edge of the Mt. Evans Wilderness Area up to Guanella Pass. It offers a variety of scenery, including spectacular views of Mt. Bierstadt and the sheer rock face on the south side of Mt. Evans. The trail intersects the Abyss Lake Trail, and has a junction with Three Mile Creek Trail, thus offering interesting loop hikes. Several stretches of the trail are above treeline.

The trail is rated as "difficult" primarily due to its length and the elevation gains in traversing its entire length. Hiking the 12.2 miles from Guanella Pass down to the Deer Creek Trailhead involves 1,280 feet in elevation gains; while hiking from Deer Creek up to Guanella Pass has an elevation gain of 3,600 feet. There is a parking area at each trailhead. In addition. the Deer Creek Trailhead has a corral for horses, but remember that camping at this trailhead is prohibited. Because the trail is in a designated wilderness area, bicycles and all motorized vehicles are prohibited.

NARRATIVE:

This description of the trail is based on starting at Guanella Pass and ending at the lower terminus near Deer Creek Campground. From the Guanella Pass parking area take the trail heading south up the small hill (marked as "11990" summit on Mt. Evans Quad map). Half way up the hill a sign indicates the beginning of Trail #603. Bear left at the sign and contour east/southeast around the hill. After approximately 25 minutes of hiking, you will reach an 84 foot post cairn on your left. Bear left at the cairn and follow successive post cairns down through the willows in a southeast direction. This portion of trail is marked as "Pack Trail" on the Quad map. At mile 2.7 you will cross Scott Gomer Creek at an elevation of 11,300 ft. The trail continues southeast up to a saddle at 11,420 ft. before dropping down into an unmarked drainage east of Scott Gomer Creek. At mile 4.5 the trail meets the Abyss Lake Trail. There is a sign at this intersection. Bear right (southwest) following the arrow pointing to Deer Creek. For the nest five minutes, the Rosalie Trail follows the Abyss Lake Trail. When you reach the second sign bear left (southeast) towards Deer Creek. This is the point where the Rosalie Trail branches off from the Abyss Lake Trail. When heading down Rosalie Trail follow the arrows to Deer Creek. When hiking up from Deer Creek follow the arrows to Gaunella Pass.

The next 2.4 miles of the trail are steep in places, gaining 1.060 feet of elevation up to Rosalie Pass. This section of the trail has some spectacular views of Mt. Bierstadt and the sheer south flank of Mt. Evans. A sign marks top of Rosalie Pass. As you begin down from the Pass you will reach the upper end of Three Mile Creek Trail 0.3 miles east of the sign at the Pass. The junction is marked by a post cairn and sign. The Rosalie Trail continues in a southeasterly direction down the Deer Creek drainage, ending at the lower trailhead. The distance from Rosalie Pass to the Deer Creek Trailhead is 5.3 miles.

TRAIL PROFILE (ONE-WAY):

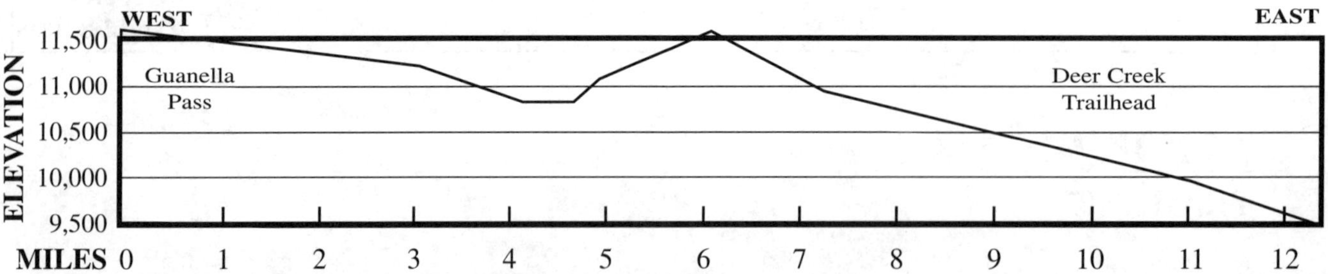

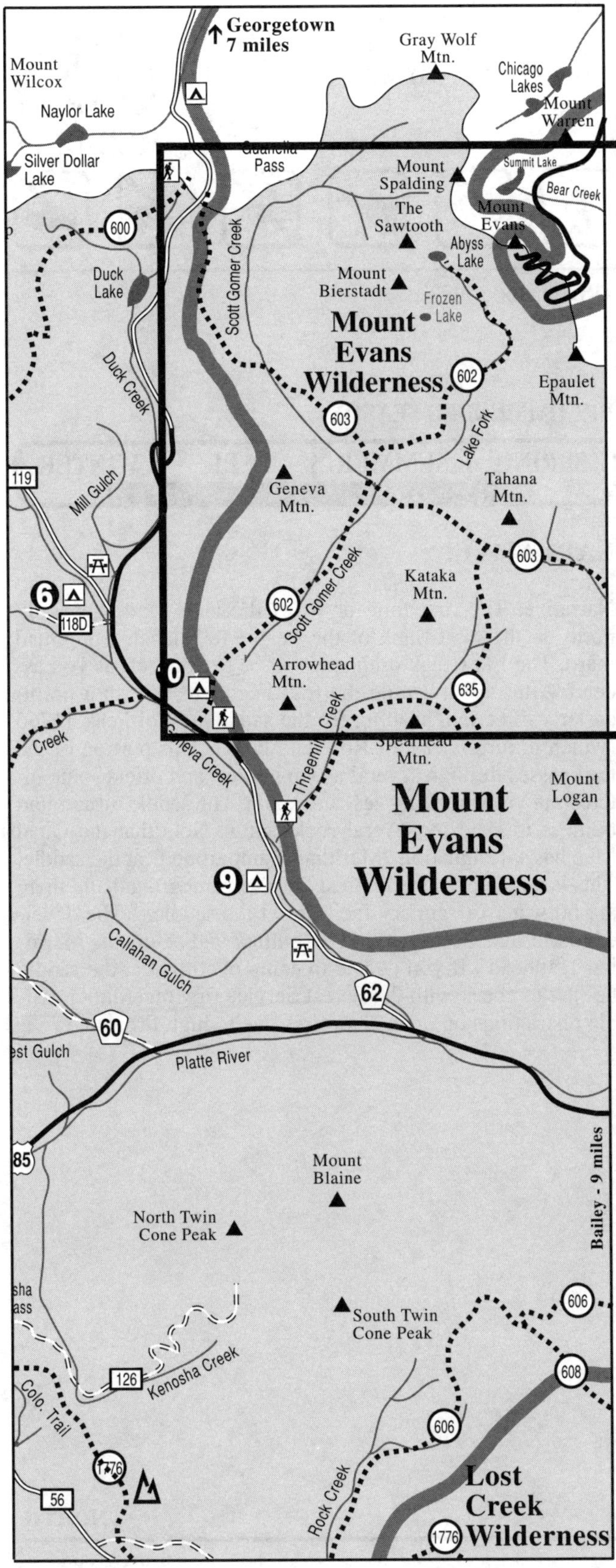

MAP 2 - East Half

TRAIL #603

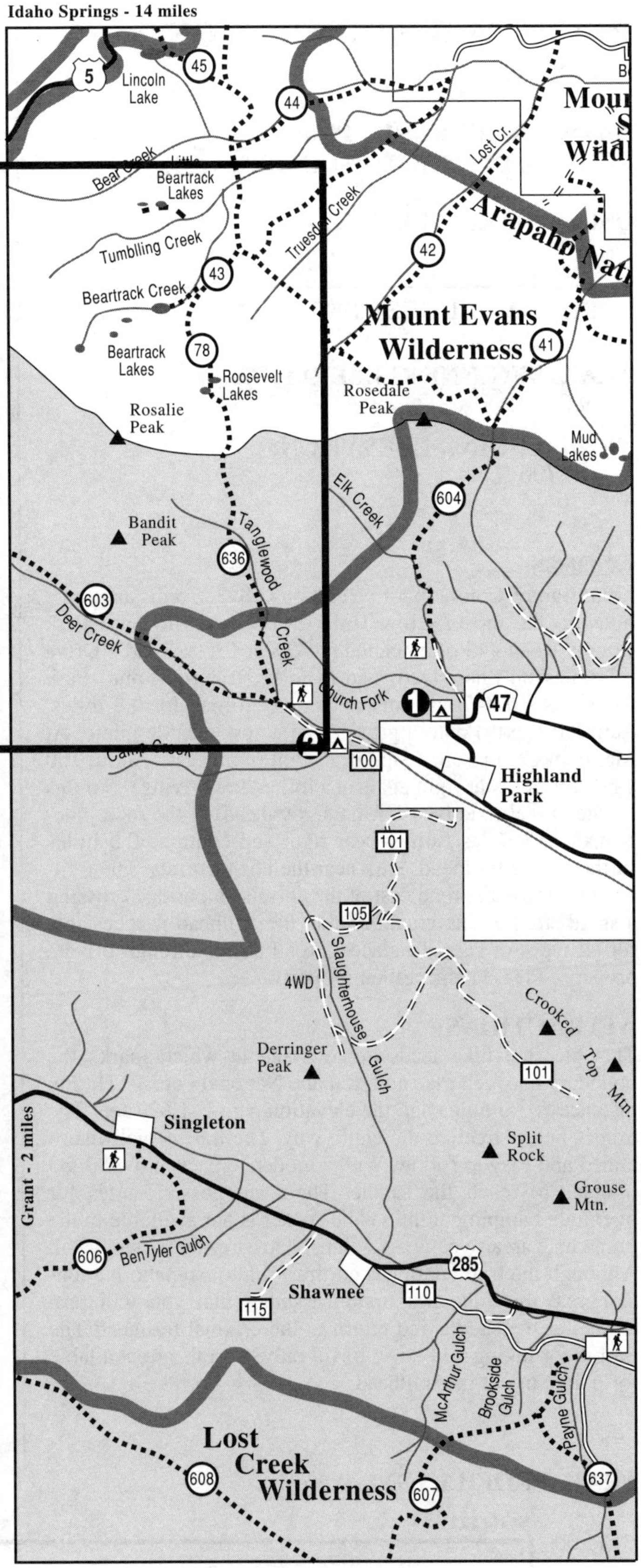

MAP 3 - West Half

ROCKY MOUNTAIN REGION
MAP NUMBER(S): 3

NATIONAL FOREST: PIKE
RANGER DISTRICT: SOUTH PLATTE

USGS: HARRIS PARK QUAD

DIFFICULTY: EASY

DISTANCE: 3.0 MILES

TRAIL 604 MERIDIAN

UPDATED: AUGUST 1990

USE: Moderate

TRAIL BEGINNING ELEVATION:
9,070 ft.

TRAIL ENDING ELEVATION:
10,700 ft.

RECOMMENDED SEASON:

SPRING	SUMMER	FALL	WINTER

ACCESS:

From Bailey, Colorado, drive east on U.S. 285 a distance of 7 miles to the top of Crow Hill. Turn left (north) onto Park County Road #43 often called the "Deer Creek Road". Drive 6.7 miles on this paved road to fork. Bear right onto Park County Road #47 and follow this paved road for 0.9 miles. Turn left (north) onto a gravel road across a cattle guard. At Meridian Campground sign bear right (northeast). About 100 yards further bear right again at a fork. After driving 0.5 miles on the gravel road bear left (straight ahead) at the road junction. At 0.3 miles further bear right and continue 0.5 miles further to the trailhead. Park near the horse corrals. The trailhead is across the road east of the corrals at a bridge crossing a small stream. The gravel road to the trailhead is accessible for all types of cars. This access road passes through private property. Please respect their property.

ATTRACTIONS:

This pleasant hike leads up to a saddle which marks the boundary between Pike and Arapaho National Forests. Hiking distance is 3.0 miles and the elevation gain is 1,630 feet The trail is below treeline the entire way. The trail is well maintained and easy to follow. With a moderate pace allow 80-9O minutes to reach the saddle. There are possible sites for overnight camping at the saddle. Water is not available in the immediate area so campers will need to carry their own water. Although the trail continues northward into Arapaho National Forest this hike description assumes that you will turn around at the saddle and return to the original trailhead The trail is for hiking and horse travel only. Corrals are available for public use at the trailhead.

NARRATIVE:

Narrative: The first mile or so of the hike heads generally north on the west flank of the Peak 9489 (labeled on Quad map). The Elk Creek drainage will be to your left as you ascend. After about a mile the trail begins to bear in a northeasterly direction, heading for the saddle east of Peak 11766 (which in turn is east of Rosedale Peak). This portion of the trail passes through several aspen groves, and offers some interesting vistas looking east and south. The saddle turnaround point is marked by several rock cairns. Note that the Quad Map has a designation "Meridian Campground" at the saddle. This is not a designated Forest Service campground and there are no water or sanitary facilities at the saddle. (The developed Meridian Campground is south of and below the Meridian Trailhead.) If you decide to camp overnight at the saddle be sure to check with the Forest Service first regarding possible restrictions on open campfires due to high fire danger.

TRAIL PROFILE (ONE-WAY):

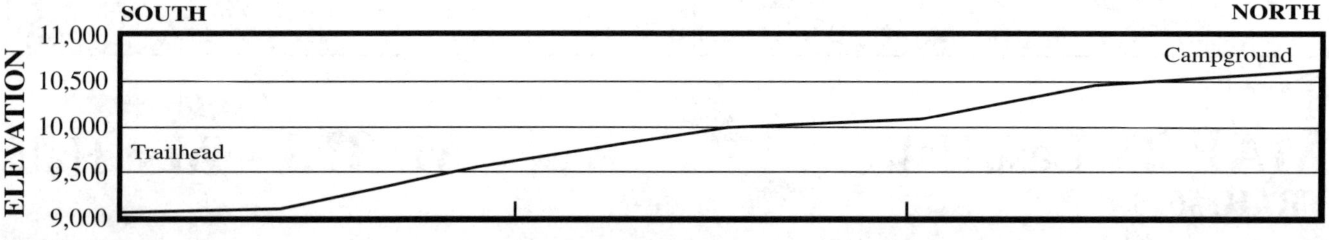

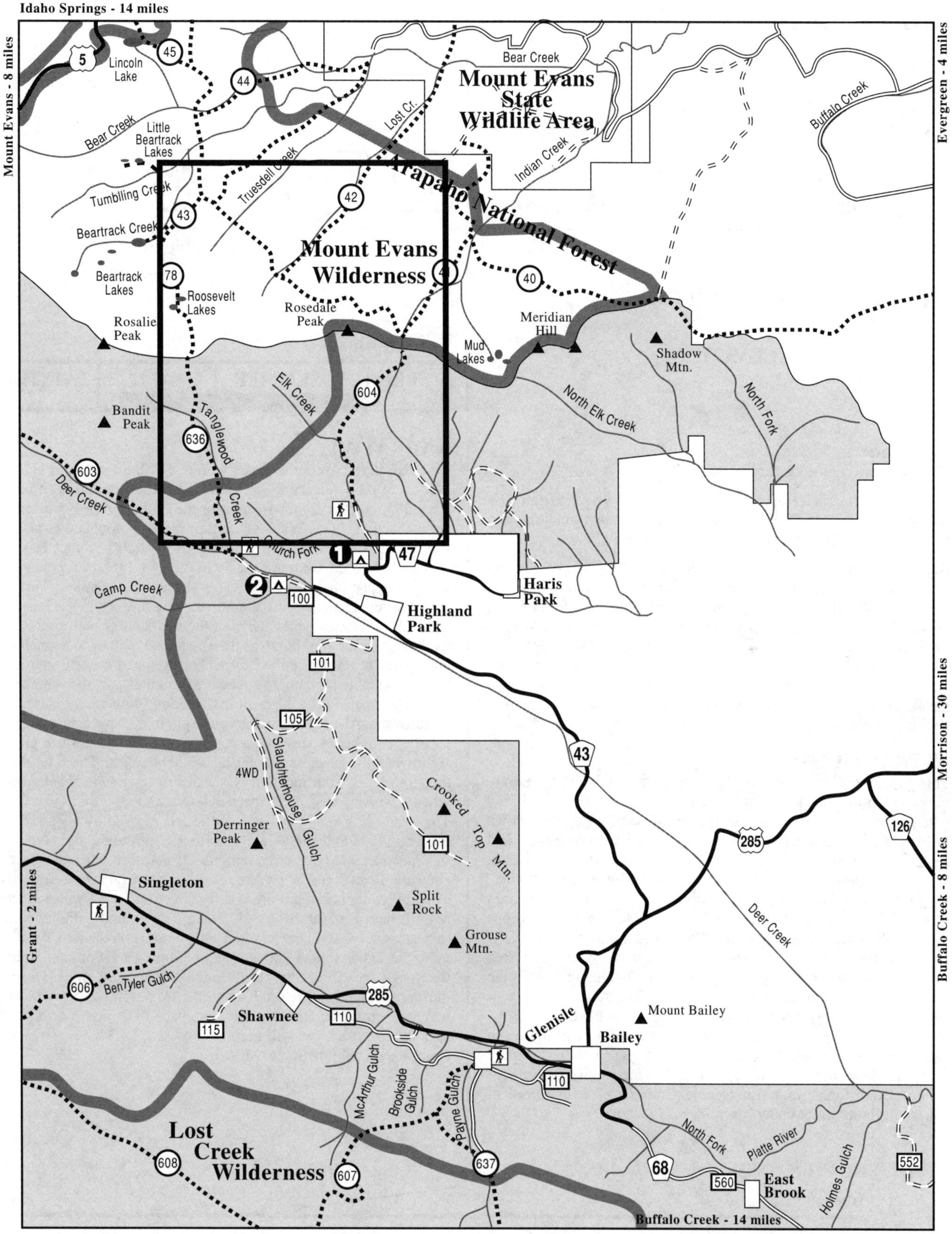

MAP 3

TRAIL #604

ROCKY MOUNTAIN REGION

MAP NUMBER(S): 2 and 3

NATIONAL FOREST: PIKE
RANGER DISTRICT: SOUTH PLATTE

USGS: SHAWNEE, MT. LOGAN, OBSERVATORY ROCK QUADS

DIFFICULTY: MODERATE TO DIFFICULT
DISTANCE: 9.0 MILES

TRAIL 606
BEN TYLER

UPDATED: AUGUST 1990

USE: Moderate, Motorized vehicles prohibited

TRAIL BEGINNING ELEVATION:
8,270 ft.

TRAIL ENDING ELEVATION:
9,720 ft.

RECOMMENDED SEASON:

SPRING | SUMMER | FALL | WINTER

ACCESS:

There are two trailheads for the Ben Tyler Trail:

Access 1. Northern Terminus, U.S. Hwy 285: Drive west from Bailey on U.S. Hwy 285 for 6.6 miles. The trailhead an a small parking area will be on the left (south) side of the highway.

Access 2.Southern Terminus, Rock Creek Cow Camp: Approximately 2.5 miles west of Kenosha Pass on U.S Highway 285 turn east on the "Lost Park Road" which is designated Forest Service Road #127. Travel about 7 miles and turn left (north) on Forest Service Road #133 - the "Rock Creek Road". Drive almost to the end of the road. Where the road forks stay to the left. The road continues to Rock Creek, but the final portion is rough with little room to turn around. It is recommended that you park above and walk down to the stream. The trail begins on the west side of Rock Creek.

ATTRACTIONS:

This popular trail is named for Ben Tyler, who lived with his family in the gulch that bears his name. He operated a lumber mill during gold rush days, hauling the sawed timber up over the ridge and down to Fairplay. The trail begins along U.S. Hwy 28 west of Bailey, and extends in a southwest direction along the west edge of the Lost Creek Wilderness Area. Five miles from the highway the trail intersects the west end of the Craig Park Trail #608. Past this trail junction it continues up over a saddle in the Kenosha Mountains, then drops down to the southern terminus of the trail near the Rock Creek Cow Camp. While the trail itself is entirely outside of the Wilderness Area you can enter the Lost Creek Wilderness Area by hiking east on the Craig Park Trail from its junction on this trail. The trail is open to foot and horse travel only, and all of the trail lies below treeline.
Major attractions include the numerous and colorful wildflowers in late spring and early summer. The trail also offers spectacular scenery of the mountain ranges to the north and the changing colors in September. Don't forget your fishing pole. Craig Creek is an excellent fishery for brook trout under seven inches in length.

NARRATIVE:

This narrative covers the first five miles of the trail from U.S.Hwy 285 up to the Craig Park Trail junction. From the trailhead of U.S. Hwy 285 the trail climbs steeply for the first 0.4 miles in a series of eleven switch backs. When you return please do not cut across these switch backs. After the last switch back the trail becomes less steep up to a stream crossing at Mile 1. At this point the trail gradient becomes steeper, climbing 1800 feet in the next three miles. There is a sign located 1.1 miles past the stream crossing saying "Craig Park Trail 1.9 miles". Just past the sign there is a short path on the right leading from the trail down to the stream and across to a small open area on the north side of the stream. This is one of the few areas along the trail suitable for camping. About 35 minutes past the sign the trail begins to open into a high alpine meadow with many varieties of wildflowers. Also, at this point you begin to get spectacular views of a large, continuous stand of aspen to the west and north a glorious sight when the aspen turn color in late September. The final mile to the Craig Park Trail junction is a series of long switchbacks which, when traversing east, offer spectacular views down Ben Tyler Gulch. The Craig Park Trail junction at Mile 5 is marked by a sign indicating Craig Park to the left (east) and Rock Creek to the right (west). At this junction you have three options. If you turn left (east) you can enter the Lost Creek Wilderness Area on the Craig Park Trail #608. The trail continues in a southeasterly direction for 5.8 miles at which point it intersects the Brookside Trail. The second option would be to bear right (west) at the junction and continue on the Ben Tyler Trail up to the saddle (about 700 feet of additional elevation) and down to the Rock Creek Cow Camp trailhead, a distance of 4 miles. Or, as the third option, you can return back to the U. S. Hwy 285 trailhead the way you came. Based on a moderate pace of 2 miles per hour plus periodic rest stops you will need to allow 2 1/2 - 3 hours to reach the Craig Park Trail junction at Mile 5.

TRAIL PROFILE (ONE-WAY):

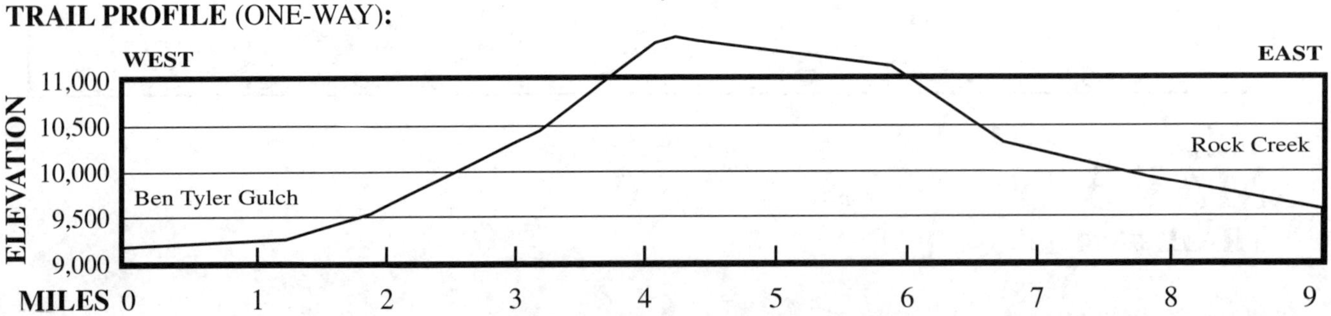

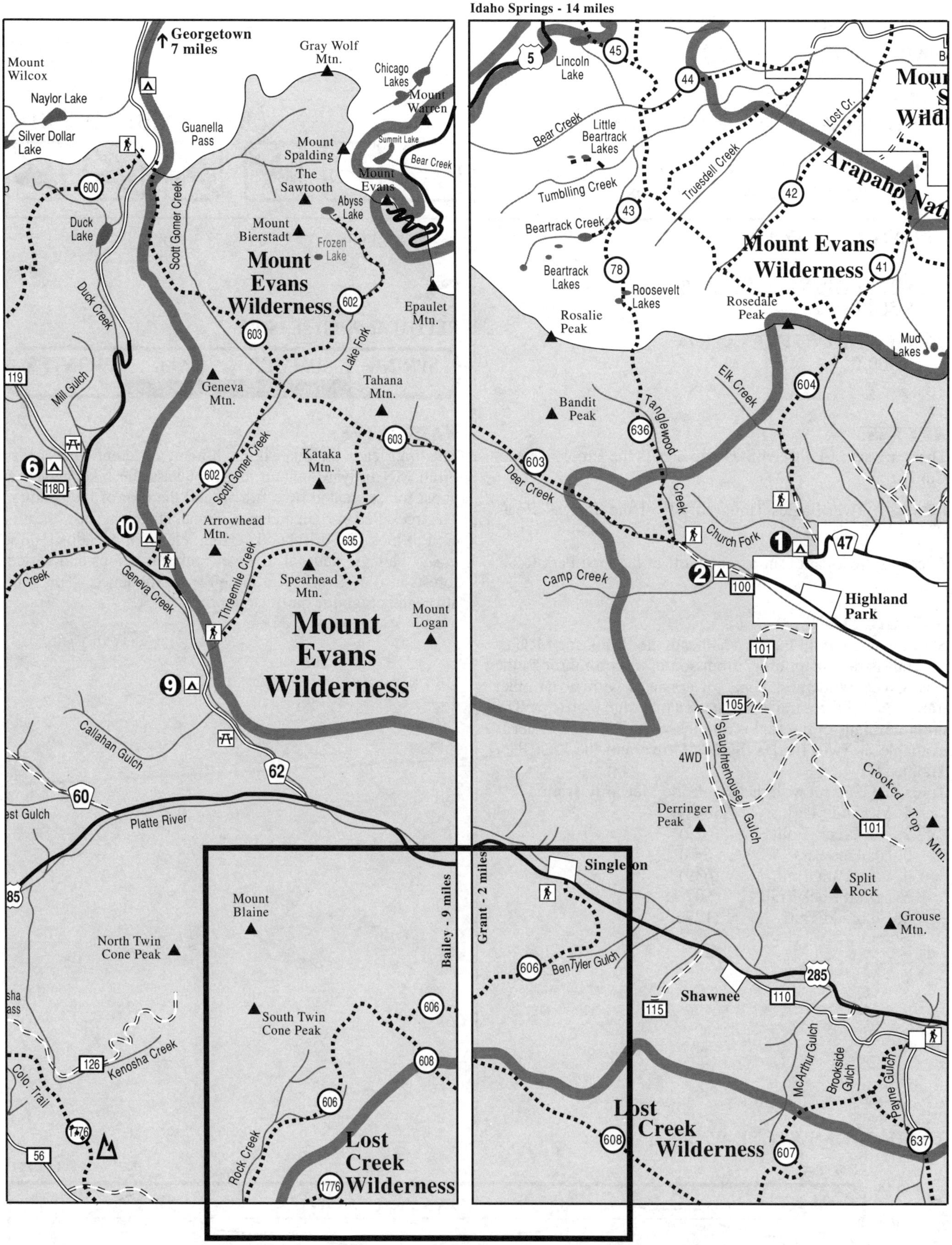

MAP 2 - East Half

TRAIL #606

MAP 3 - West Half

ROCKY MOUNTAIN REGION

MAP NUMBER(S): 3 and 9

NATIONAL FOREST: PIKE
RANGER DISTRICT: SOUTH PARK

USGS: McCURDY MOUNTAIN., FARNUM PEAK, TOPAZ, SHAWNEE QUADS

DIFFICULTY: MODERATE TO DIFFICULT
DISTANCE: 37 MILES

TRAIL 607
BROOKSIDE McCURDY

UPDATED: MAY 1989

USE: Light

TRAIL BEGINNING ELEVATION:
8,500 ft.

TRAIL ENDING ELEVATION:
9,900 ft.

RECOMMENDED SEASON:

SPRING | SUMMER | FALL | WINTER

ACCESS:
There are several different ways to access the Brookside McCurdy trail;

Access 1. Twin Eagles Trailhead, off of the Tarryall Road. (Co. Rd. 77).

Access 2. Lost Park Campground. off of the Lost Park Road. (FDR 127).

ATTRACTIONS:
Starting at the Twin Eagles Trailhead, the Brookside McCurdy Trail is the "main drag" from south to north through the Lost Creek Wilderness Area. It connects with seven other trails, most of these trails can be used to design loop trips. The north terminus of the trail is 3 miles west of Bailey. Water is available at Twin Eagles, Spruce Grove and the Lost Park Trailheads.

Trails that connect with the Brookside-McCurdy Trail:

1. McCurdy Park (628)
2. Ute Creek Trail (629)
3. Hankins Pass (630)
4. Lake Park Trail (639)
5. Lizard Rock Trail (607.1)
6. Colorado Trail (1776)

NARRATIVE:
This trail leads the hiker through many different landscapes. From McCurdy Mountain to Bison Pass the hiker enters a ghost forest created by a fire around the turn of the century. The trees died upright and many remain that way, now weathered to look like driftwood. High elevation meadows low grassy valleys unusual granite rock formations and varied forests also add variety to the hiker's experience. This trail also enters bighorn sheep country.

TRAIL PROFILE (ONE-WAY):

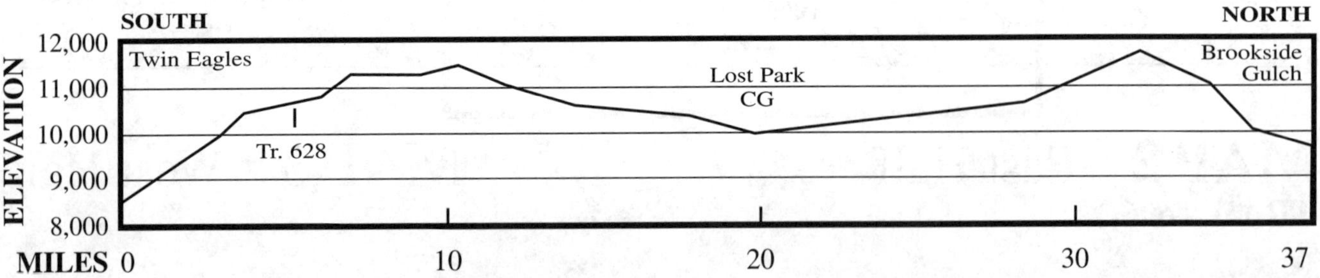

MAP 3 - South Fifth

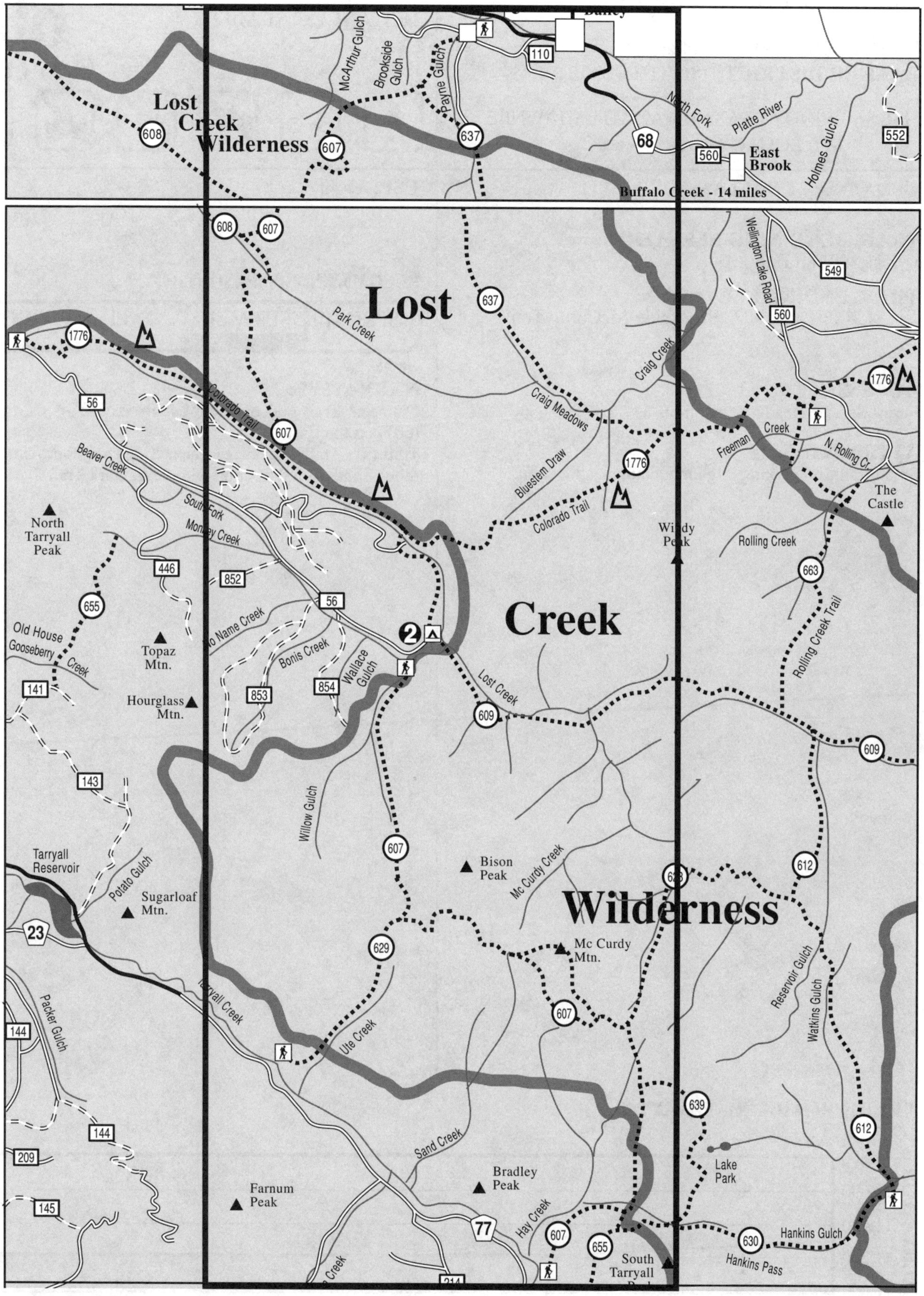

MAP 9 TRAIL #607

ROCKY MOUNTAIN REGION

MAP NUMBER(S): 2, 3, and 9

NATIONAL FOREST: PIKE
RANGER DISTRICT: SOUTH PLATTE

USGS: WINDY PEAK, TOPAZ MT., SHAWNEE MT. LOGAN QUADS

UPDATED: UNKNOWN

TRAIL BEGINNING ELEVATION:
11,300 ft. (Tr. 606)

TRAIL ENDING ELEVATION:
11,200 ft. (Tr. 607) Brookside-McCurdy Trail

ACCESS:
Forest Service Road 543 and trails 637, 607 and 606

ATTRACTIONS:
Small trailhead parking, high mountain meadow trail.

DIFFICULTY: MODERATE

DISTANCE: 6.0 MILES

TRAIL 608
CRAIG PARK

USE: Light

RECOMMENDED SEASON:

SPRING	SUMMER	FALL	WINTER

NARRATIVE:
Connects Trail 606 on the West Brookside McCurdy Trail (607) on the East
Trail 606 Trailhead located near Singleton-approximately 4 miles South to intersection with Craig Park Trail

TRAIL PROFILE (ONE-WAY):

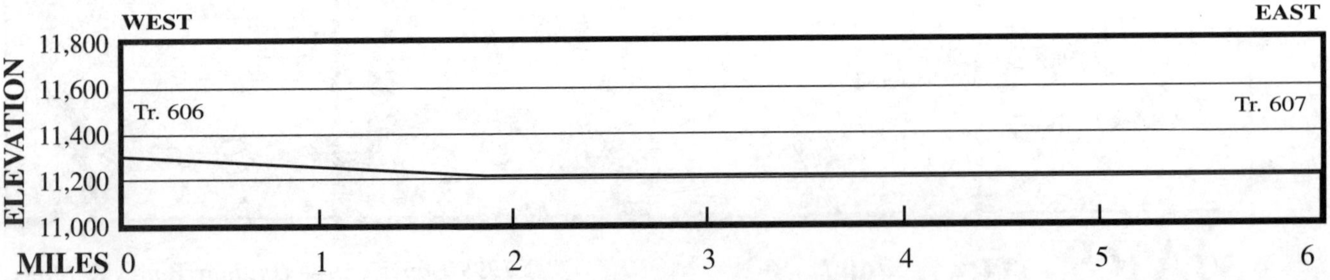

MAP 2 - Southeast Quarter

MAP 3 - Southwest Quarter

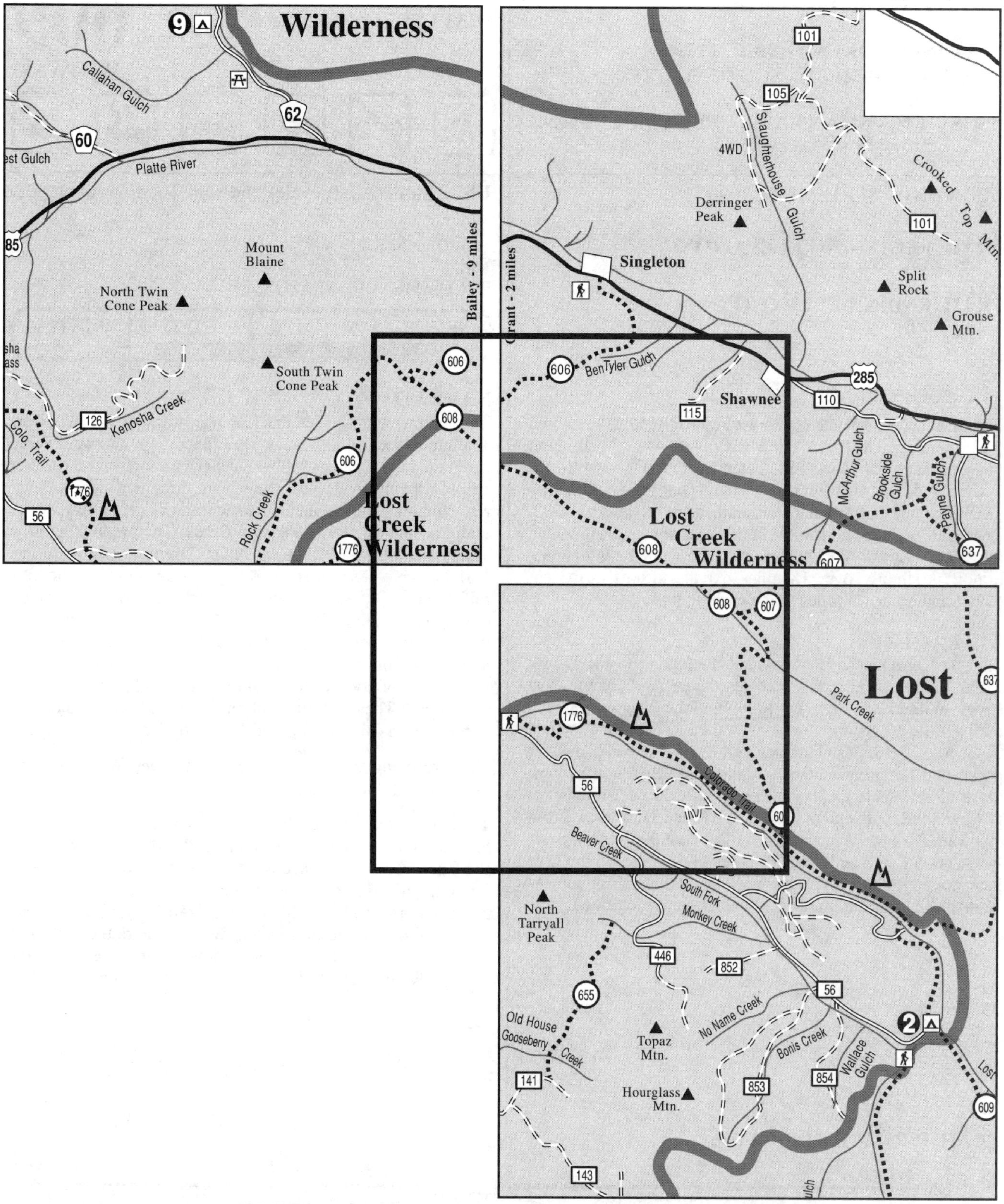

TRAIL #608

MAP 9 - Northwest Quarter

ROCKY MOUNTAIN REGION
MAP NUMBER(S): 9 and 10

NATIONAL FOREST: PIKE
RANGER DISTRICT: SOUTH PLATTE

USGS: CHEESMAN LAKE, GREEN MOUNTAIN, WINDY PEAK QUADS

DIFFICULTY: MODERATE

DISTANCE: 10.8 MILES

TRAIL 609 WIGWAM

UPDATED: SEPTEMBER 1990

USE: Moderate, Bicycles and vehicles prohibited

TRAIL BEGINNING ELEVATION:
8,160 ft.

TRAIL ENDING ELEVATION:
9,500 ft.

RECOMMENDED SEASON:

SPRING	SUMMER	FALL	WINTER

ACCESS:

From Bailey, head South on Forest Service Road #543, which is also identified as Park County Road #68. At the Wellington Lake entrance F.S. Road #543 bears left (east) toward Buffalo Creek-take the right fork instead and continue south across the Wellington Lake dam. (This south-bound road is now F.S. Road #560) Continue 6 more miles to a sign indicating the turnoff to Wigwam Trail. Total distance from Bailey to this turnoff is slightly over 18 miles. At the turnoff turn right (west) and drive 1.3 miles to the parking lots.

ATTRACTIONS:

This trail provides a less crowded alternative to the Goose Creek Trailhead. The trail accesses the east side of the Lost Creek Wilderness Area by following the Wigwam Creek drainage up to an open meadow called "Wigwam Park"- a distance of 3.5 miles. The meadow has a number of beaver ponds and flat areas suitable for camping. In Wigwam Park the trail intersects the north end of the Goose Creek Trail #612 and the south end of the Rolling Creek Trail #663. From Wigwam Park the Wigwam Trail itself continues in a westerly direction for an additional distance of 6.6 miles to the Lost Park Campground. Since the trail lies within a Wilderness Area all bicycles and motorized vehicles are prohibited.

NARRATIVE:

This narrative describes the first 4.2 miles of the Wigwam Trail up to the Rolling Creek Trail junction in Wigwam Park. From the parking areas follow the signs a short distance down an old logging road. Near the stream there is a sign on your right marking the beginning of the trail. The trail heads generally northwest along Wigwam Creek. The grade is a fairly steady rise of 1,340 ft. in 3.5 miles. The last mile up to the Park has several short, steep, rocky sections of trail. After 3.0 miles you will pass by a sheer vertical rock wall that comes down to the right edge of the trail. At mile 3.5 you will reach the Gooses Creek Trail junction in Wigwam Park. The Wigwam Trail continues in a westerly direction along the north side of beaver ponds. The Rolling Creek Trail junction sign is at mile 4.2. These 2 trail junctions are clearly marked on the "Photo revised 1984" copy of the Windy Peak Quad Map.

Upon reaching the open meadow area called Wigwam Park you have 4 hiking options:

1. Return back to the trailhead the way you came.
2. Head south on the Goose Creek Trail #612. The distance to the Goose Creek Trailhead is 9.6 miles. Most of this trail will be on the McCurdy Mt. Quad Map.
3. Head north on the Rolling Creek Trail #663. The trailhead is about 6.4 miles north. Leaving Wigwam Park the trail rises steeply to 10,650 ft., then drops down a very steep section into the Rolling Creek drainage. Although not marked on the Windy Peak Quad Map, the trailhead is in Section 24 just west of F.S. Road #543.

TRAIL PROFILE (ONE-WAY):

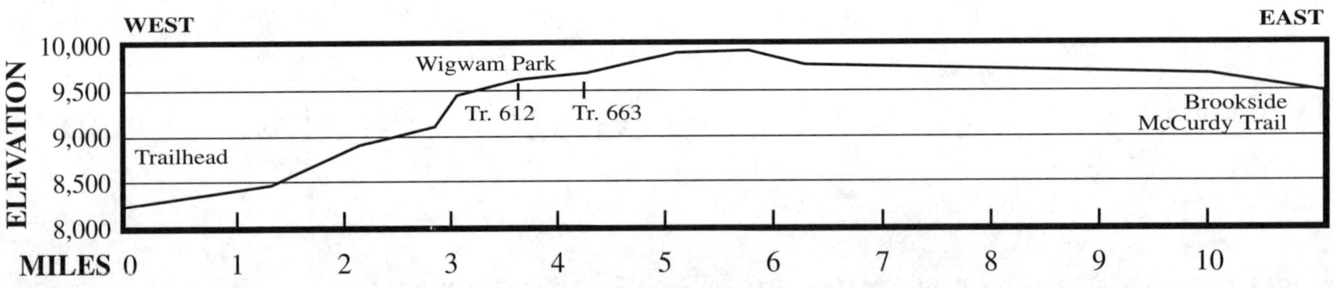

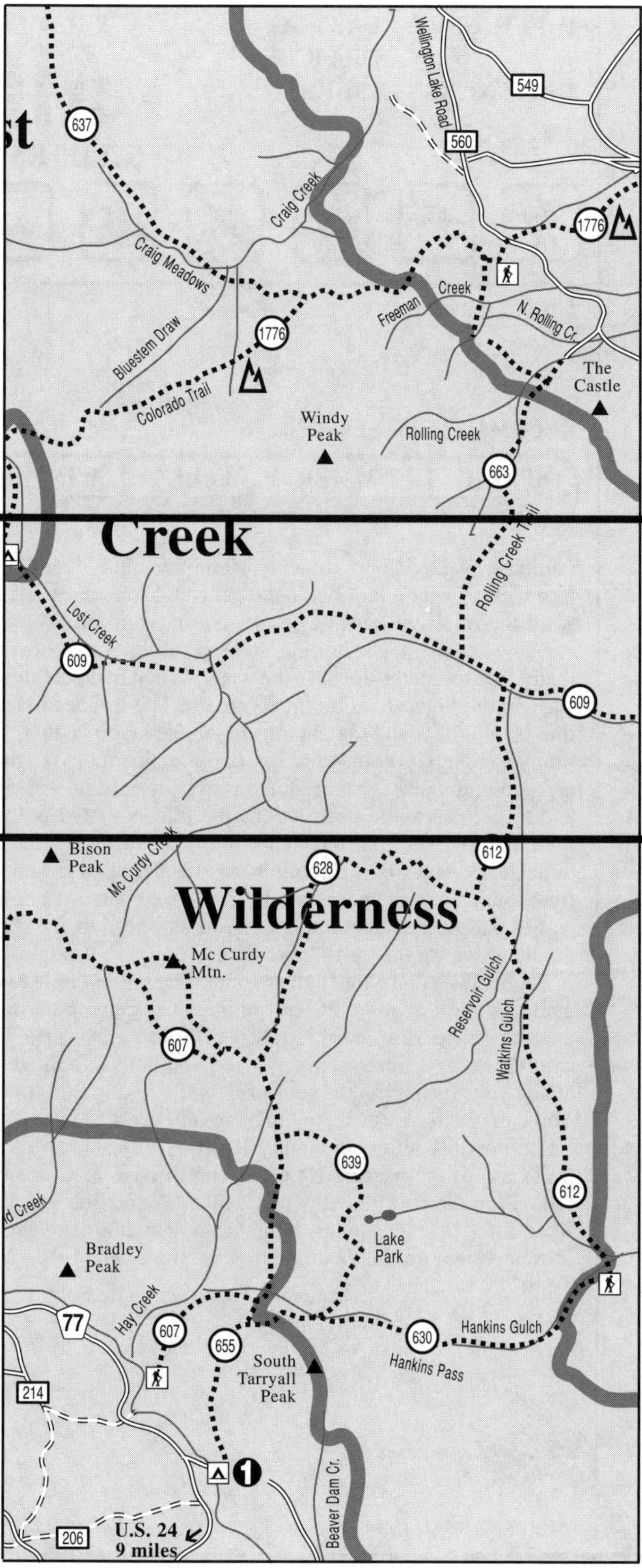

MAP 9 - East Half

TRAIL #609

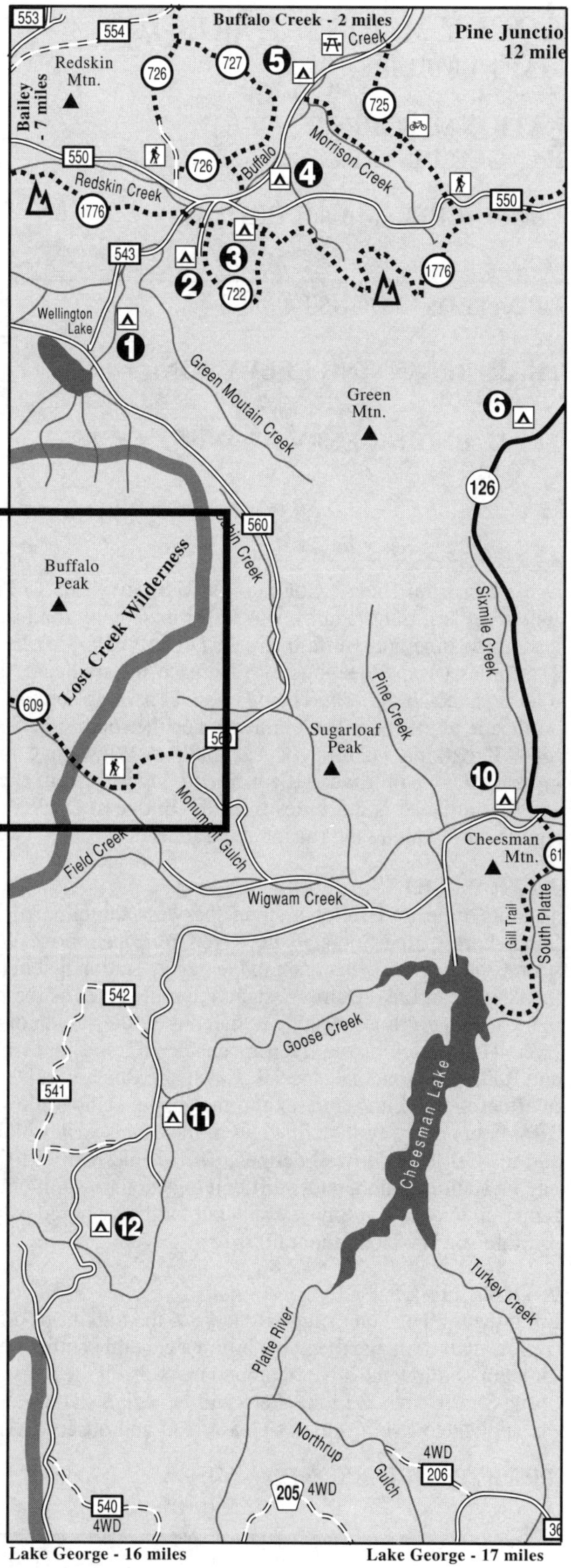

MAP 10 - West Half

ROCKY MOUNTAIN REGION
MAP NUMBER(S): 10

NATIONAL FOREST: PIKE
RANGER DISTRICT: SOUTH PLATTE

USGS: CHEESMAN LAKE QUAD

DIFFICULTY: EASY TO DIFFICULT
DISTANCE: 3.5 MILES

TRAIL 610
GILL TRAIL

UPDATED: AUGUST 1990

USE: Heavy

TRAIL BEGINNING ELEVATION:
6,548 ft.

TRAIL ENDING ELEVATION:
6,610 ft.

RECOMMENDED SEASON:

SPRING	SUMMER	FALL	WINTER

ACCESS:
Drive east from Bailey, Colorado on U.S. Hwy 285 to Pine Junction. Turn right (south) onto Jefferson County road #126 leading to Pine and Buffalo Creek. Drive on Hwy #126 for 21.9 miles (from Pine Junction) to reach the trailhead. The road is paved all the way. There Is a small parking area on the north side of the road. The trailhead is on the south side of the road. If parking is full, you can park at Wigwam Campground. 1/3 of a mile west. If you are driving west from Deckers, the trailhead is 2.5 miles from the bridge at Deckers. A sign Is posted along the road at the trailhead.

ATTRACTIONS:
This 3.5 mile trail follows along the west shoreline of the South Fork of the South Platte River from just above Wigwam Club (private), upstream to the gaging station just below the Cheesman Lake Dam. As such, the trail serves as the primary access for those wishing to fish this 3 mile stretch of the river. This stretch of the river is considered good for Brown and Rainbow Trout, and special 'Catch and Release' rules are in effect. If you choose to hike to the gaging station allow 2-2 1/4 hours one way. The first half of the trail is easy walking and the trail is fairly well defined. The second half up to the gaging station is moderately difficult to negotiate, with only a faint trail to follow at times. Watch out for the scattered patches of poison ivy along the trail.

NARRATIVE:
From the trailhead on County Road #126 the trail drops down over a small drainage then up a hill in a general southeasterly direction. Follow the silver diamond markers on the trees and posts. Shortly into the hike there will be a sign stating all of the applicable fishing rules - please read and observe. After you have walked approximately 10 minutes, the river comes into view to your left down in the canyon. From this point onward the river will always be in view on your left. The trail stays well above the waterline, so be careful not to follow the many side footpaths down to the water. When in doubt always stay on the highest visible trail or path. The trailhead elevation is 6548 feet and the gaging station elevation is 6610 feet - only a slight elevation gain. But don't let this fool you. there are numerous places where huge rocks jut out into the river and to get past these obstructions the trail is routed up over the outcrops. The trail has many 'ups and downs', some are quite steep. The first 11/2 mile is easy walking on a well defined trail. Then as you pass mile 2, the trail narrows and becomes almost indistinct in places. You will need to always be on the watch for the trail - 'hike with your eyes'. Again. when in doubt always take the highest route. The last 3/4 mile to the gaging station is quite difficult hiking. The canyon narrows, the slope increases and the trail is all loose gravel. Be very careful here! At times you may need both hands to move, so attach your fishing rod to your pack while hiking this stretch. You can reverse this hike by driving on Forest Service Road #211 (turn off Jefferson county Road #126 just west of the trailhead) to the parking lot on the north edge of Cheesman Lake. You have to hike down a Denver Water Board Access Road for 1 1/4 miles to reach the gauging station. The gaging station shown on the Quad map is the upper end of the Gill Trail.

TRAIL PROFILE (ONE-WAY):

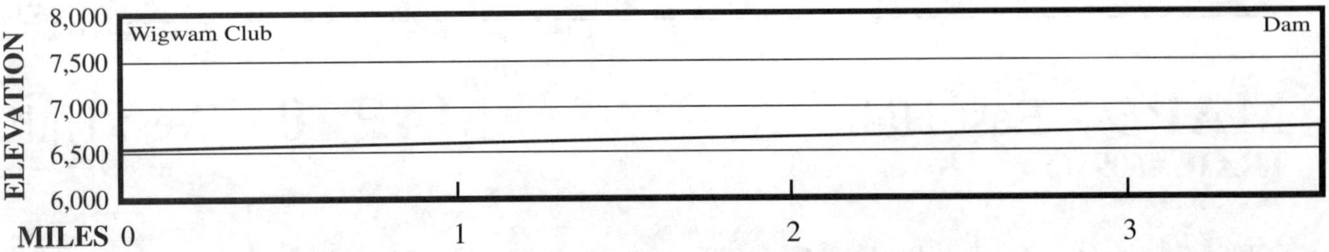

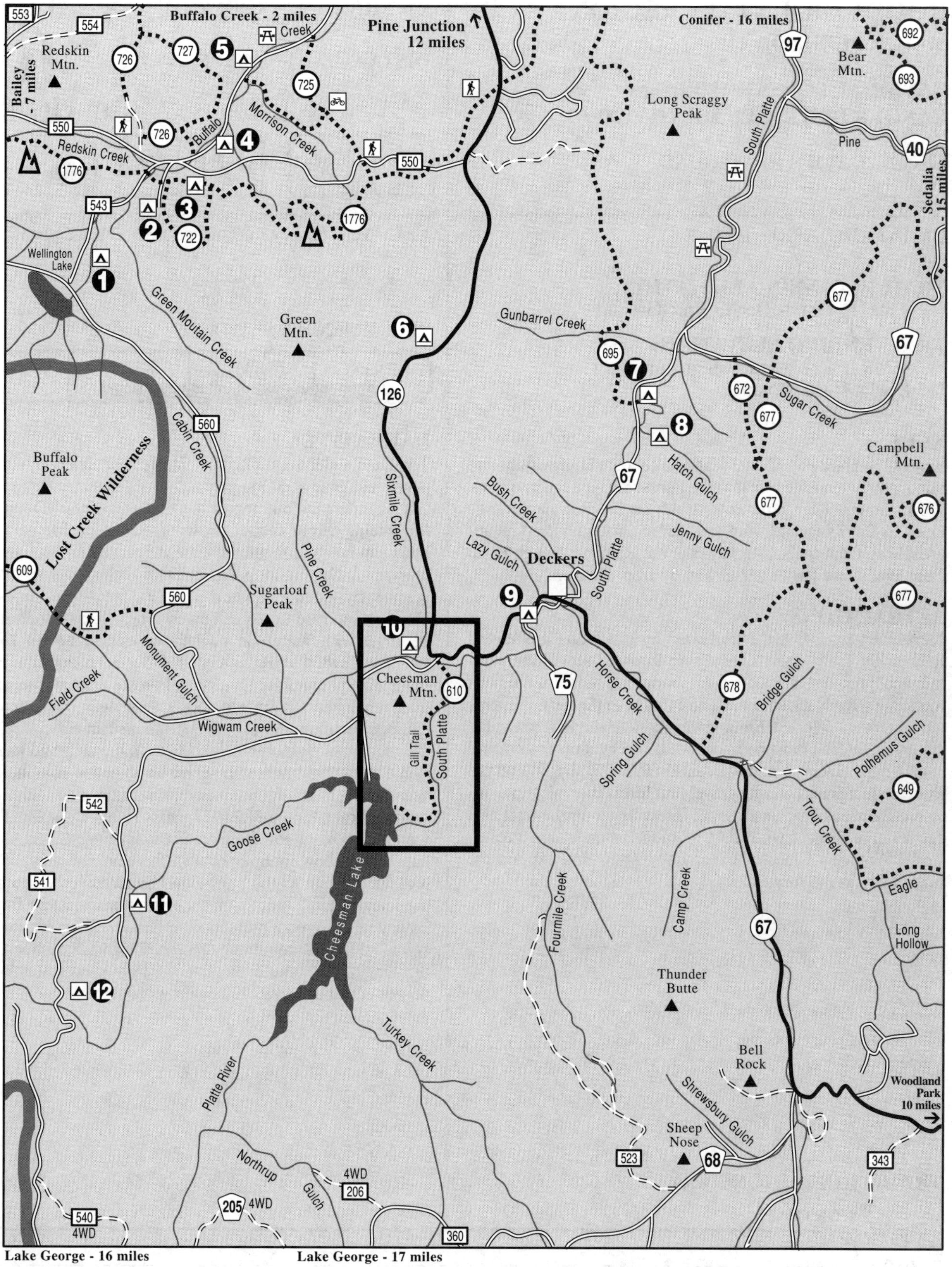

MAP 10

TRAIL #610

ROCKY MOUNTAIN REGION

MAP NUMBER(S): 11

NATIONAL FOREST: PIKE
RANGER DISTRICT: SOUTH PLATTE

USGS: DEVILS HEAD QUAD

DIFFICULTY: MODERATE

DISTANCE: 1.3 MILES

TRAIL 611
DEVILS HEAD

UPDATED: APRIL 1980

USE: Very Heavy, motorized vehicles are prohibited

TRAIL BEGINNING ELEVATION:
8,800 ft, Devils Head Picnic Ground

TRAIL ENDING ELEVATION:
9,748 ft, Lookout tower at summit of Devils Head Mountain

RECOMMENDED SEASON:

SPRING	SUMMER	FALL	WINTER

ACCESS:
From U.S. Highway 85 at Sedalia, take State Highway 67 approximately ten miles west to the Forest Service Indian Creek Work Center. Then nine miles south on the Rampart Range Road to Devils Head Camp and Picnic Grounds. Persons approaching from the south may take the Rampart Range Road from Woodland Park or Highway 67 from Deckers.

ATTRACTIONS:
November 16, 1979, the trail was designated as a National Recreation Trail. Devils Head Fire Lookout sets at the summit and provides a 360 degree panoramic view of the surrounding Pike National Forest and Plains to the east. Visitors are welcome. While Climbing the stairs to the lookout at the summit, children must be under adult supervision and control at all times. Hikes should use caution to avoid slipping on the loose disintegrated granite gravel that forms the trail tread. Interpretive signs have been set at intervals along the trail as a cooperative project with the Colorado Business and Professional Women's Clubs of Colorado to show and explain the role of fire in the forest.

NARRATIVE:
The Devils Head Trail climbs 948 feet in elevation through a ponderosa pine and Douglas-fir forest. It culminates at a massive granite rock outcrop at the highest point of Devils Head Mountain. From certain viewpoints, the profile of a devils head can be seen formed by the rugged rock outcrops of the mountain. Setting atop this massive rock pinnacle, made accessible by stairways over the last 200 feet to the summit, sets Devils Head Fire Lookout Tower. The original structure built in 1912 with materials packed up the mountain by pack mules, provided a magnificent 360 degree panoramic view of the surrounding Pike National Forest. Later, the existing glass enclosed ten by ten foot station was constructed and equipped with modern fire detection instruments. Because of the numerous lightning caused fires in the area and its exceptional vista, the tower still serves in an active role in wildfire detection. It is the last operating fire lookout along Colorado's Front Range. In 1917, Miss Helen Dow was the first woman Lookout for the Pike National Forest, and was perhaps the first woman lookout in the United States. The fire lookout is open to the public and hundreds of visitors make the journey each year. Certain restrictions must be followed, however, for your protection. Please observe the posted signs. Helpful reminders: wear comfortable shoes, bring drinking water, take camera, bring light jacket, stay on trail, do not climb the stairs if lightning is occurring.

TRAIL PROFILE (ONE-WAY):

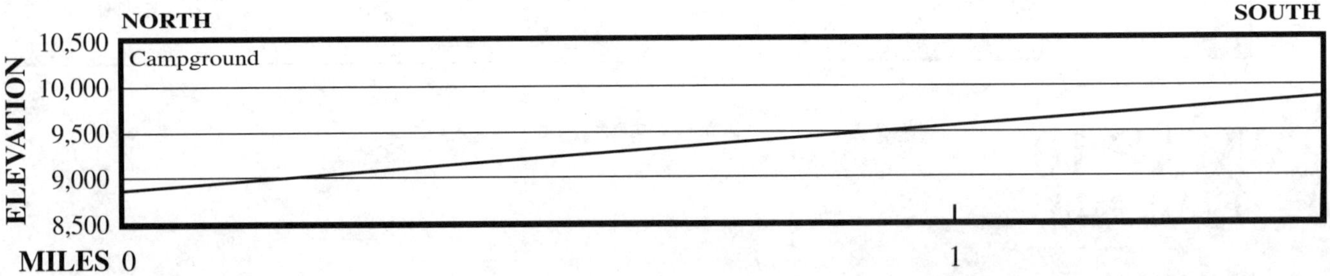

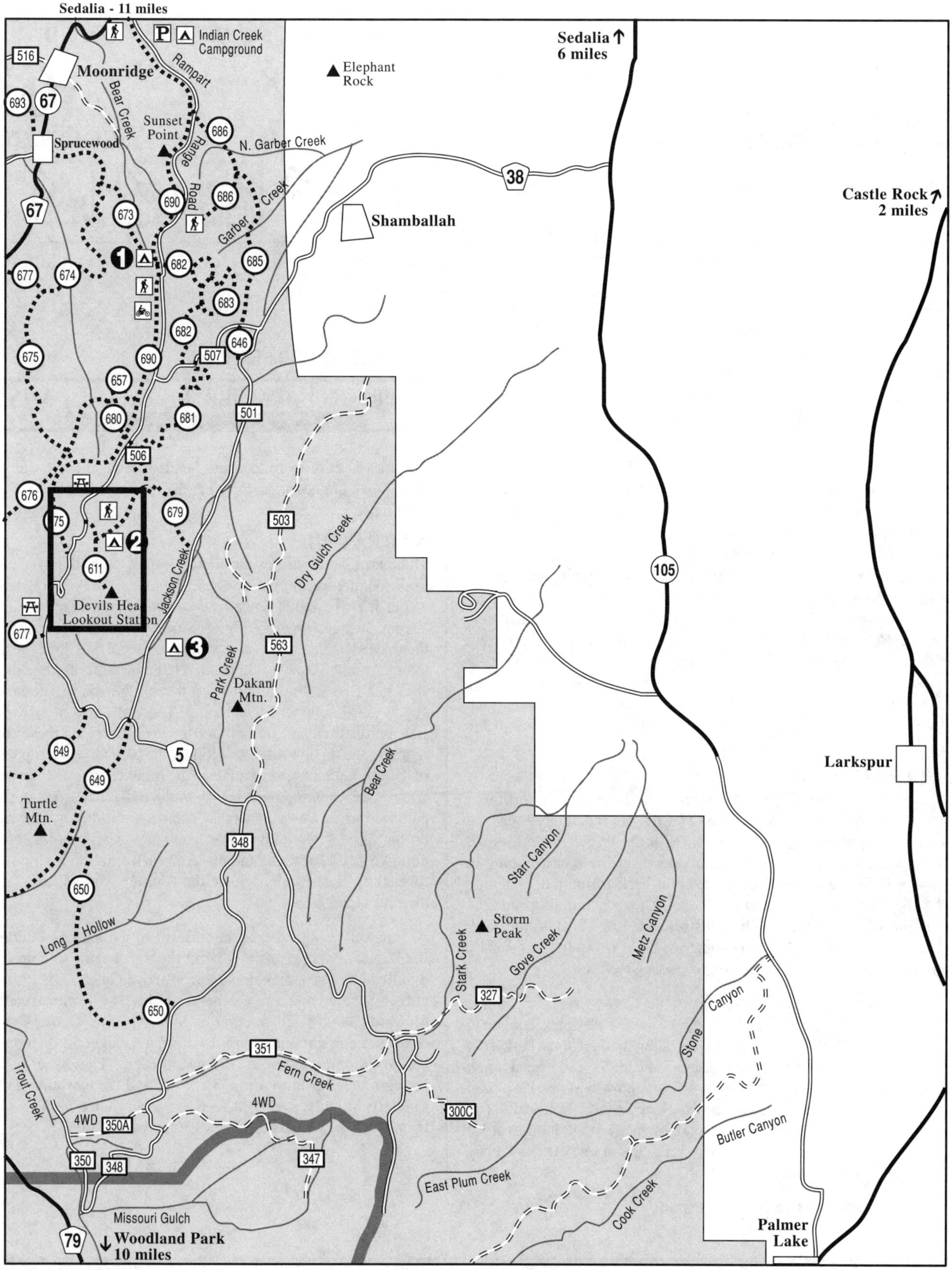

MAP 11

TRAIL #611

ROCKY MOUNTAIN REGION

MAP NUMBER(S): 9

NATIONAL FOREST: PIKE
RANGER DISTRICT: SOUTH PLATTE

USGS: McCURDY MT., WINDY PEAK QUADS

DIFFICULTY: MODERATE

DISTANCE: 9.6 MILES

TRAIL
612
GOOSE CREEK

UPDATED: SEPTEMBER 1990

TRAIL BEGINNING ELEVATION:
8,200 ft.

TRAIL ENDING ELEVATION:
9,500 ft.

USE: Moderate, Bicycles and vehicles prohibited

RECOMMENDED SEASON:

SPRING | SUMMER | FALL | WINTER

ACCESS:

From Bailey, drive east on U.S. Hwy 285 to Pine Junction. Turn right, (south) at Pine Junction onto Jefferson County Road #126 towards Pine and Buffalo Creek. Drive on County Road #126 a distance of 21.8 miles. Turn right (south) onto Forest Service road #211 that leads to Cheesman Lake. After 2 miles turn right again (west) at the sign pointing to Goose Creek. Drive 1.1 miles in westerly direction until you reach a fork in the road. Bear left at the fork and stay on F.S. Road #211. From this fork drive 5.2 miles to a road intersection just south of Molly Gulch Campground. Turn right and drive 4.7 miles to the Goose Creek Trailhead access road. About half way along this segment you will pass Goose Creek Campground. At the trailhead access sign turn right (north) and drive 1.3 miles to the trailhead parking.

ATTRACTIONS:

This very popular trail provides access into the southeast portion of the Lost Creek Wilderness Area. The trail is 9.6 miles long and ends in Wigwam Park at a junction with the Wigwam Trail #609. The Goose Creek Trail offers numerous campsite areas, fishing in Goose Creek, and vistas of spectacular rock formations. At mile 3.9 there is a side trail leading to several historic buildings and the remains of a shaft house. These historic sites date back to the early 1900's when an attempt was made to, build a water reservoir by creating an underground dam on the Lost Creek.

The McCurdy Park Trail #628 intersects the Goose Creek trail at mile 6.5. A popular back pack "loop" is to hike the Goose Creek Trail up to the McCurdy Park Trail, turn left and follow the McCurdy Park Trail to its intersection with the Brookside Trail #607. Then head south about 1 mile to the Lake Park Trail #639. Follow the Lake Park Trail to Hankins Pass, then hike east on the Hankins Pass Trail #630 back to the Goose Creek Trailhead. The total distance of this "loop' is about 24 miles. This entire loop is shown on the 'Photo revised 1984' copy of the McCurdy Mountain Quad Map.

The trail is easy to follow and heads in a general northerly direction. There are horse corrals near the trailhead that are available for public use.

NARRATIVE:

This narrative describes the first 6.5 miles up to the McCurdy Park Trail junction. From the trailhead sign near the parking area follow a spur trail 0.2 miles down to a wooden foot bridge across a small stream. Just across the bridge is a trail junction. Hankins Pass Trail #630 heads left (west) and Goose Creek Trail #612 heads right (east). At mile 0.6 you will reach a large metal footbridge that crosses Goose Creek. Ten minutes past this bridge you will come to open, grassy areas suitable for camping. There are also numerous campsites on the opposite side of Goose Creek. To reach these campsites return to the metal bridge and follow the path along the opposite side of Goose Creek. Approximately 20 minutes past the metal bridge you will reach an unmarked fork in the trail. The left fork is a small path leading down to the Creek. The main trail follows the right fork up the hill. After hiking 7 or 8 minutes past the fork look to the northwest and locate a "thumb-like" rock spire up high on the rock formation. Just to the right of the "thumb" you will see an arch in the rocks.

At mile 3.19 a sign marks the turn off to the historic buildings and shafthouse remains. At this point the trail begins to gain elevation rapidly before dropping down into Watkins Gulch. The trail then begins to head up again in a general northwest direction, reaching the McCurdy Park Trail junction at Mile 6.5. The Goose Creek Trail continues straight ahead (north) up to a saddle east of :Peak 10459 before dropping down into Wigwam Park. The distance from the junction to Wigwam Park is 3.1 miles and the elevation gain to the saddle is slightly over 700 feet above the junction. The McCurdy Park Trail #628 heads left (west) from the junction.

TRAIL PROFILE (ONE-WAY):

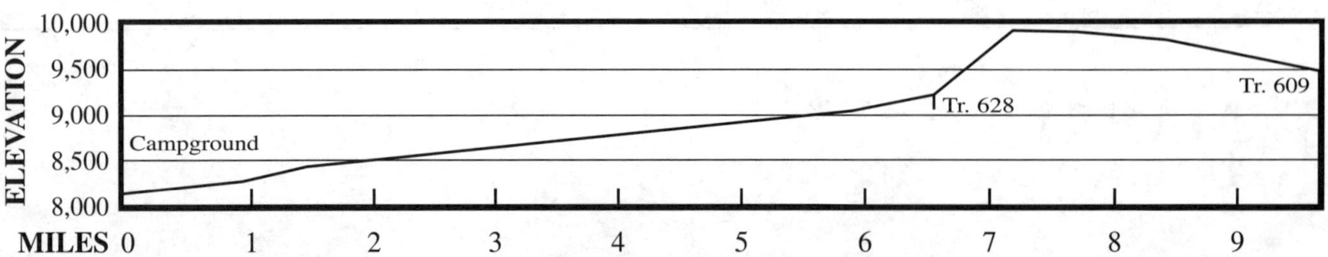

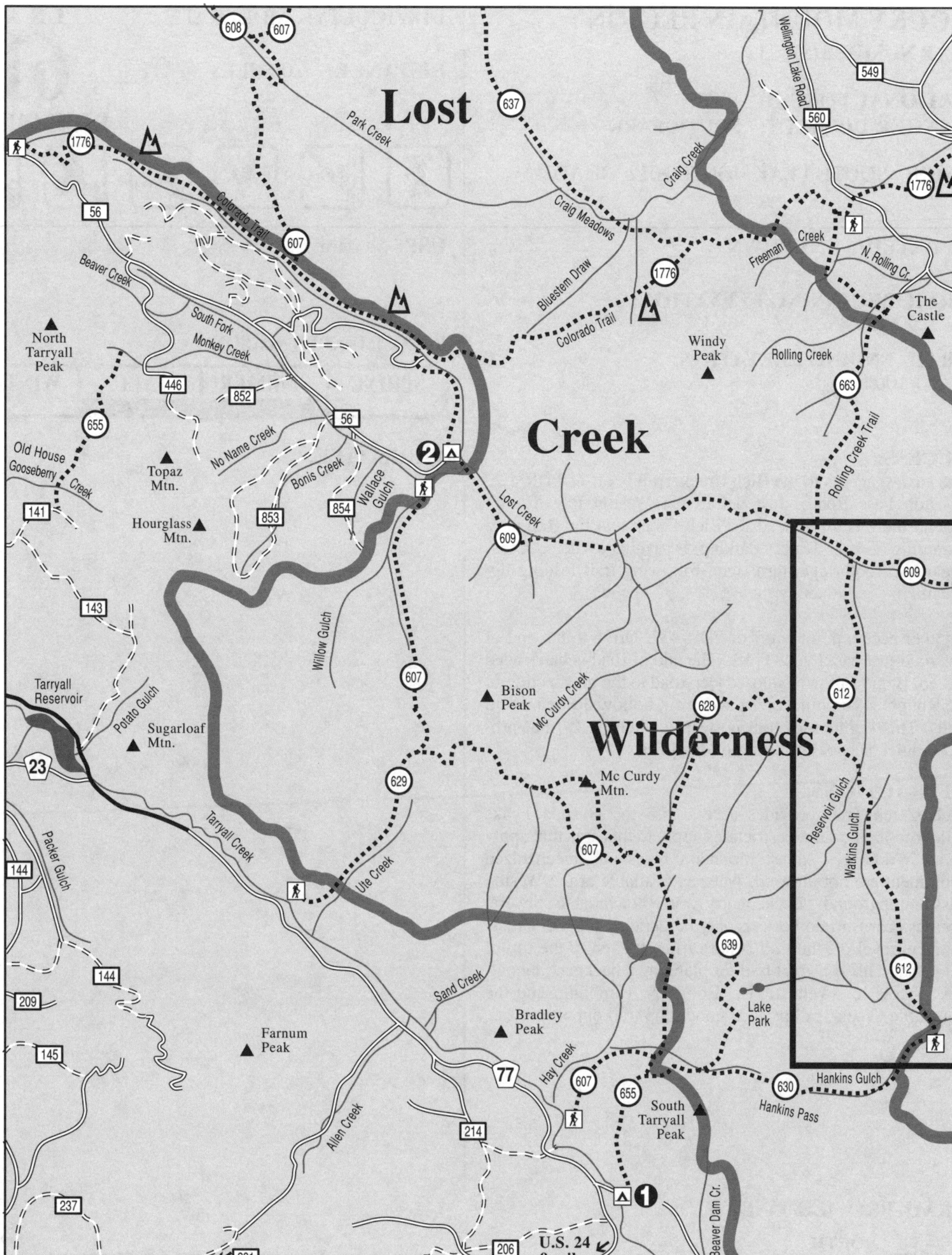

MAP 9

TRAIL #612

ROCKY MOUNTAIN REGION

MAP NUMBER(S): 12

NATIONAL FOREST: PIKE
RANGER DISTRICT: SOUTH PARK

USGS: SOUTH PEAK, JONES HILL QUADS

DIFFICULTY: MODERATE

DISTANCE: 6.0 MILES

TRAIL
616
RICH CREEK

UPDATED: UNKNOWN

TRAIL BEGINNING ELEVATION:
9,900 ft.

TRAIL ENDING ELEVATION:
11,100 ft.

ACCESS:
The easiest access to the Rich Creek Trail is off of FDR 425. (Weston Pass Road). Just before the Weston Pass Campground there is a trailhead for Rich Creek and the Rough & Tumbling Creek Trails. Parking is available here. Cross stream on footbridge, turn right, following trail leading upstream.

Another access point is off of FDR 431. Drive to the end of this road just past Lynch creek. The end of road is barricaded closed. Begin following closed jeep road to the intersection of the Rough & Tumbling creek and Trail. Follow upstream trail (left). This segment of trail is steep (switch backs) but provides quick access into Buffalo Meadows.

ATTRACTIONS:
Rich Creek Trail provides access into the Buffalo Peaks Wilderness. The same restrictions apply to this area that apply to all Wilderness Areas. Mountain bikes and mechanized equipment are not allowed. Water is available at the Weston Pass Campground. Backcountry water is available. Several streams near Buffalo Peaks is only moderately rugged. Cross-country travel off the trail is not difficult, even to the top of the peaks. This is a great trail for planning a backpacking outing. Access to several trails; Rough & Tumbling, and the Four-mile Creek Trail provide good loop trail opportunities.

USE: Moderate

RECOMMENDED SEASON:

SPRING	SUMMER	FALL	WINTER

NARRATIVE:

TRAIL PROFILE (ONE-WAY):

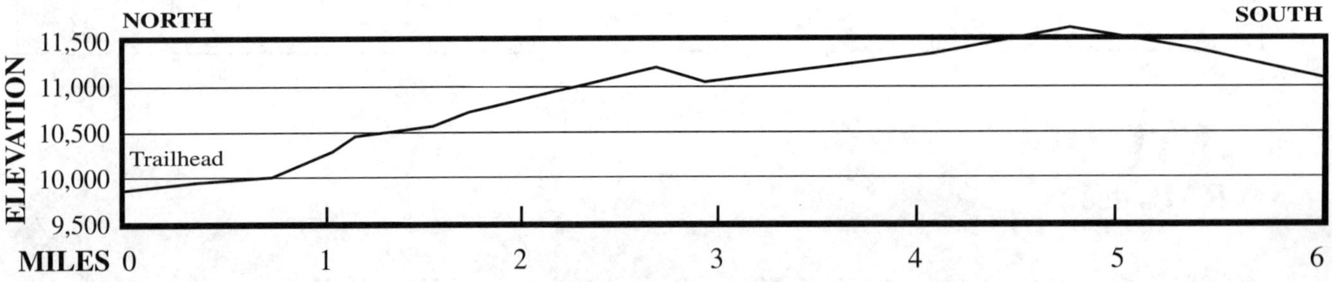

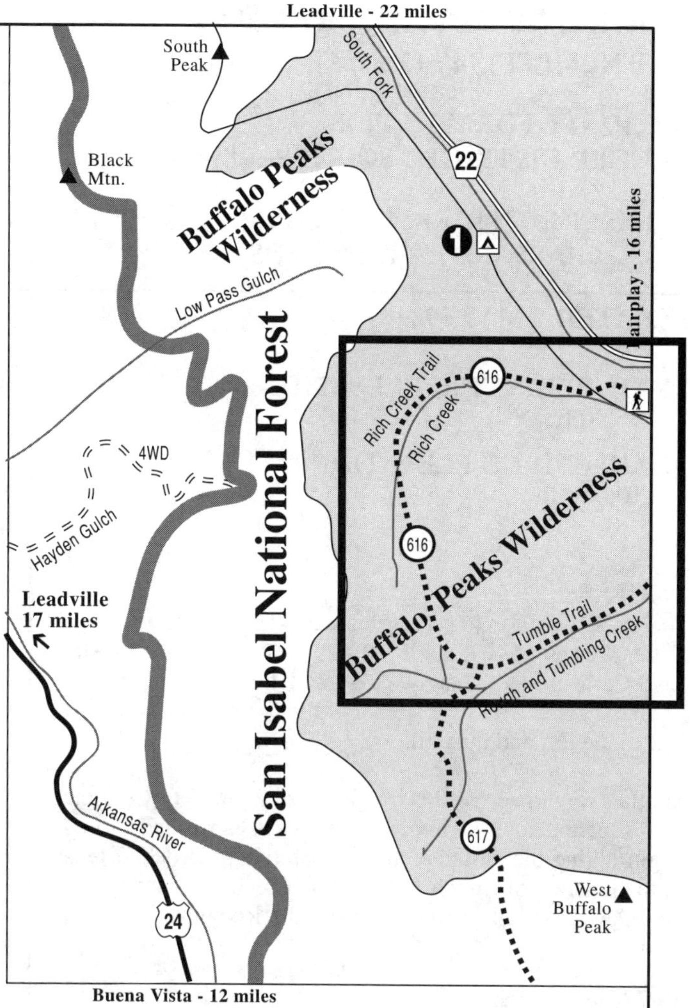

MAP 12
TRAIL #616

ROCKY MOUNTAIN REGION

MAP NUMBER(S): 12 and 13
San Isabel 6
NATIONAL FOREST: PIKE
RANGER DISTRICT: SOUTH PARK

USGS: SOUTH PARK, JONES HILL, HARVARD LAKES QUADS

DIFFICULTY: MODERATE

DISTANCE: 9.0 MILES

TRAIL 617

ROUGH & TUMBLING CREEK

UPDATED: MAY 1989

TRAIL BEGINNING ELEVATION:
9,900 ft.

TRAIL ENDING ELEVATION:
10.000 ft.

ACCESS:

There are two ways to access this trail. The easiest way is off of the Weston Pass Road, FDR 425. Just before the Weston Pass Campground, there is a signed trailhead. Parking is available here. Cross the stream of the footbridge and follow trail to the left and up a hill.

Another way to access this trail is off of FDR 431. Continue to end of the road. This road has been barricaded closed. Begin hiking the closed jeep road paralleling Lynch -Creek. At the junction of Rough & Tumbling Creek and the trail you can hike either left or right. This is the Rough & Tumbling Trail.

ATTRACTIONS:

The Rough & Tumbling Trail gives you access into the Buffalo Peaks Wilderness. The same restrictions apply to this area that apply to all Wilderness Areas. This is a great trail for planning a backpacking trip. Access with the Four-mile Creek Trail and the Rich Creek Trail provide good loop trail opportunities. Water is available at the Weston Pass Campground. Backcountry water is also available.

USE: Moderate. Mountain bikes and other mechanized equipment prohibited

RECOMMENDED SEASON:

SPRING	SUMMER	FALL	WINTER

NARRATIVE:

TRAIL PROFILE (ONE-WAY):

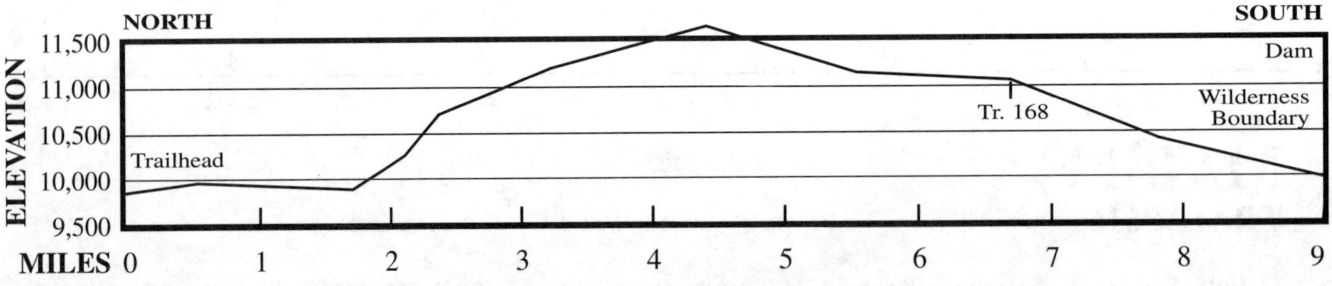

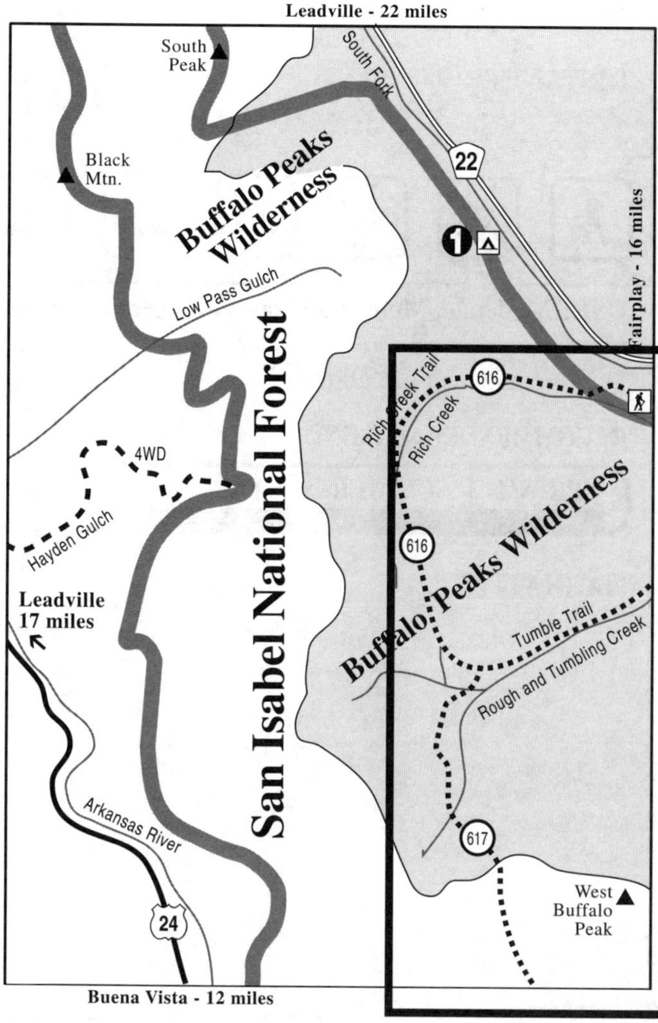

MAP 12 - Northwest Quarter

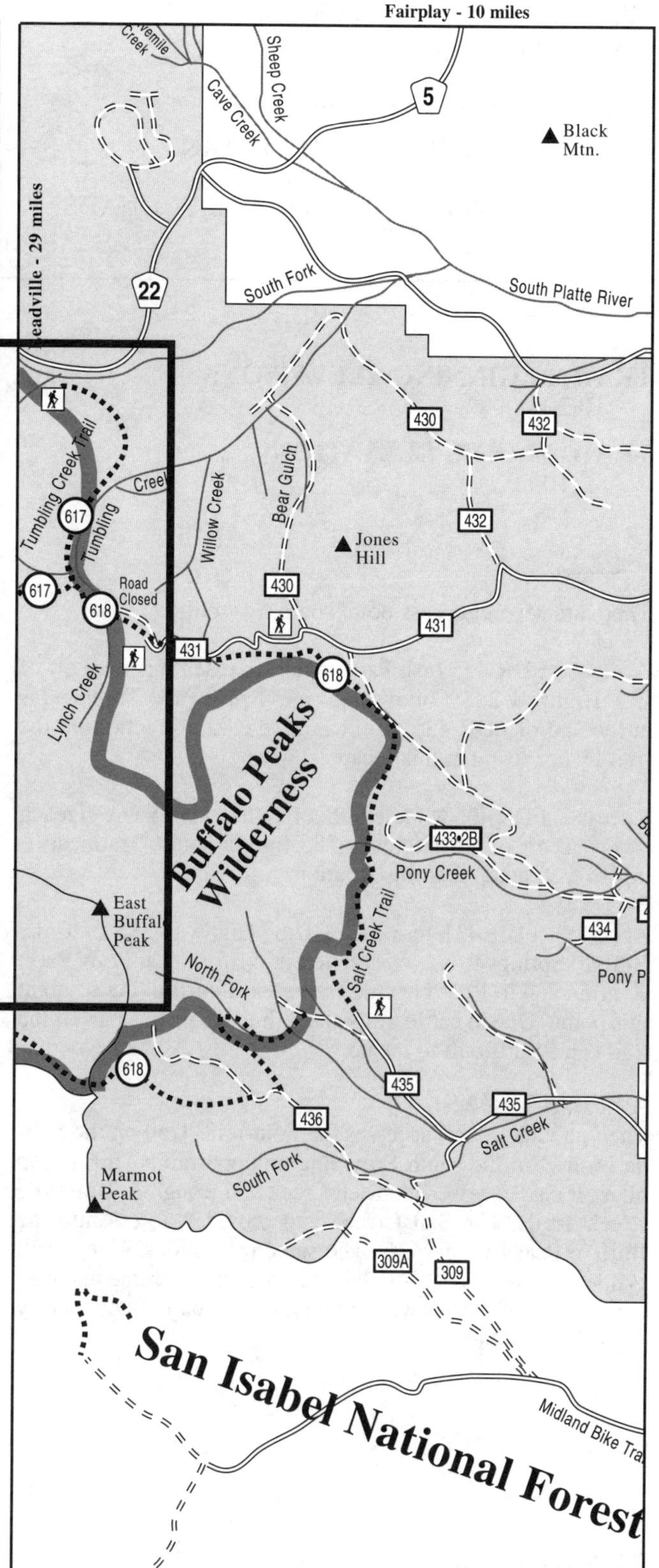

MAP 13 - West Half

TRAIL #617 (San Isabel National Forest)

ROCKY MOUNTAIN REGION

MAP NUMBER(S): 13
San Isabel 6
NATIONAL FOREST: PIKE
RANGER DISTRICT: SOUTH PARK

USGS: JONES HILL, MARMOT PEAK QUADS

DIFFICULTY: MODERATE

DISTANCE: 7.0 MILES

TRAIL 618
SALT CREEK

UPDATED: MAY 1989

TRAIL BEGINNING ELEVATION:
10,200 ft.

TRAIL ENDING ELEVATION:
10,100 ft.

ACCESS:
There are several access points onto this trail;

Access 1. FDR 435 rush Park. You can reach this road off of U.5 Highway 285. Follow signs into Brush Park. Trailhead is at the end of FDR 435. You can hike either direction on the trail by accessing at this location.

Access 2. FDR 436 Middle Fork of Salt Creek. You can reach this road off of US Highway 285 further south. Trailhead is not as well marked at this location.

Access 3. FDR 431 turn off of U.S Highway 285 onto the Buffalo Springs Road. Stay to left at the fork near Pony Park. Continue on to Bear Gulch. Trail comes in from the south at this point. Due to cattle trails and a timber sale this part of the trail can be difficult to locate.

ATTRACTIONS:
The Salt Creek Trail accesses the Four-mile Trail on the Salida District to the south Providing an opportunity for a loop hike. It also accesses the Rich Creek and Rough & Tumbling Creek Trails. The Salt Creek Trail provides access into the Buffalo Peaks Wilderness. The same restrictions apply to all Wilderness Areas. Mountain bikes and other mechanized equipment are not allowed within the boundary. Backcountry water is available.

USE: Moderate, Mountain bikes and other mechanized
equipment prohibited

RECOMMENDED SEASON:

SPRING	SUMMER	FALL	WINTER

NARRATIVE:

TRAIL PROFILE (ONE-WAY):

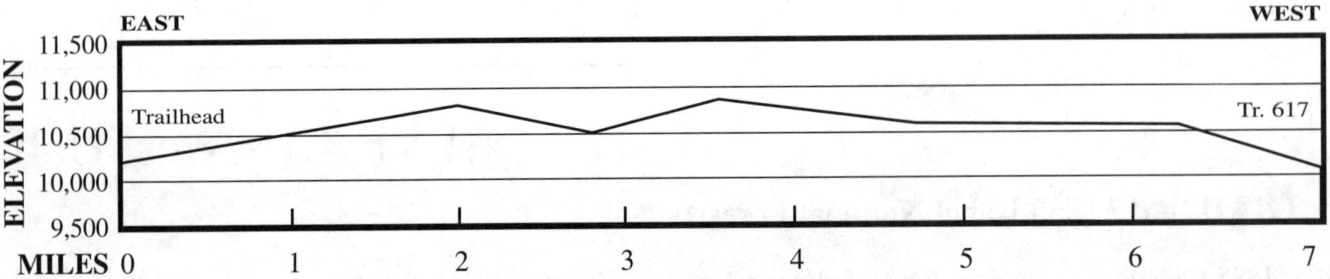

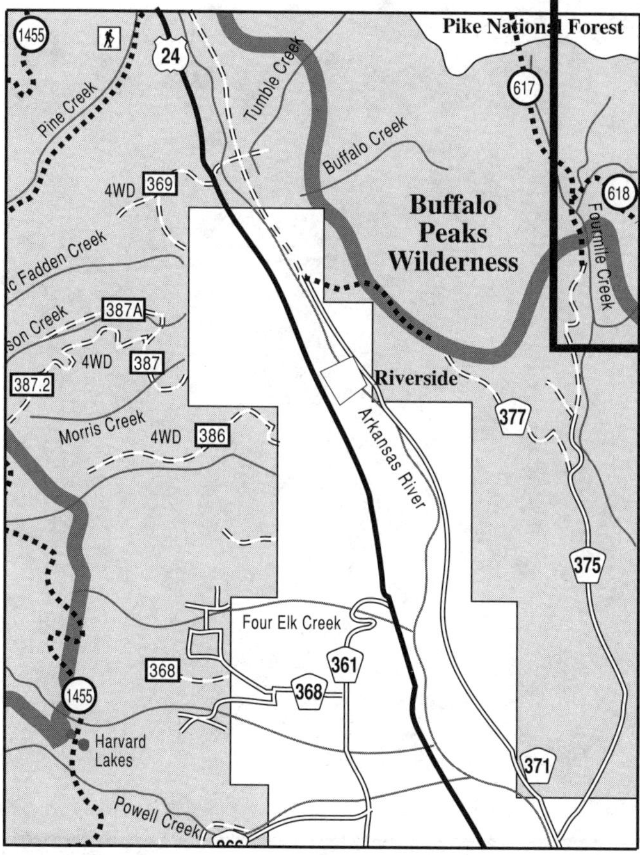

MAP 6 - Northeast Quarter

TRAIL #618 (San Isabel National Forest)

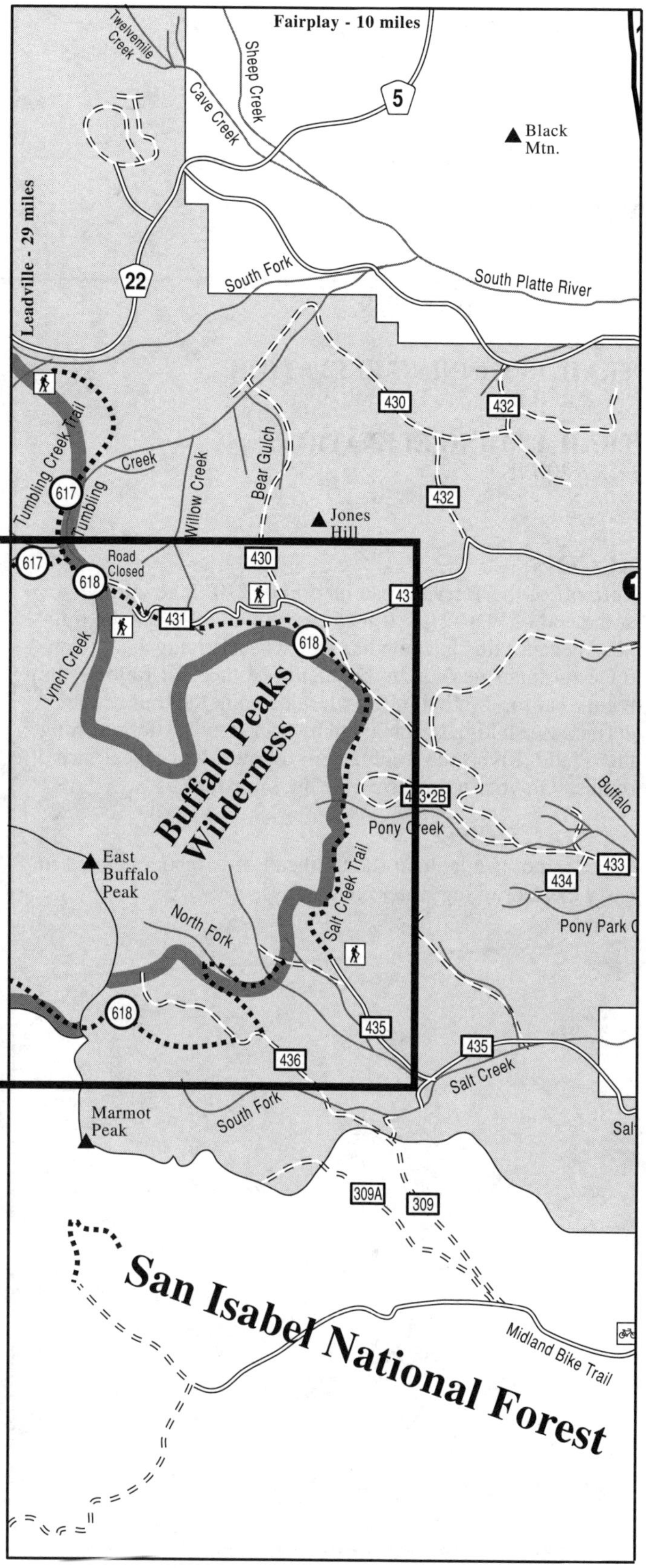

MAP 13 - West Half

ROCKY MOUNTAIN REGION

MAP NUMBER(S): 15

NATIONAL FOREST: PIKE
RANGER DISTRICT: SOUTH PARK

USGS: HACKETT MT. QUAD

DIFFICULTY: MODERATE TO DIFFICULT
DISTANCE: 3.0 MILES

TRAIL 619 LONGWATER

UPDATED: MAY 1989

TRAIL BEGINNING ELEVATION:
8,400 ft.

TRAIL ENDING ELEVATION:
7,200 ft.

ACCESS:
Turn off of the Tarryall road onto FDR 210. The trail begins at the end of FDR 210. It may be difficult to locate the trailhead because this is a popular four-wheel driving area. Continue on main road. The beginning of the trail follows the main road for the first mile. After this point the trail becomes defined to a hiking trail marked by trail signs. After reaching the Platte River the trail again follows four-wheel drive routes. On your return trip hike the same trail back.

ATTRACTIONS:
Water is not available at the trailhead. It is recommended to carry extra drinking water on hot days.

USE: Moderate

RECOMMENDED SEASON:

SPRING	SUMMER	FALL	WINTER
	■	■	

NARRATIVE:

TRAIL PROFILE (ONE-WAY):

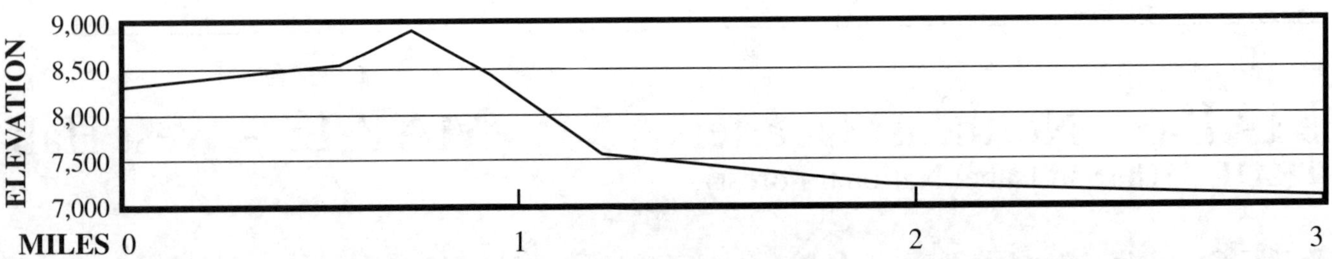

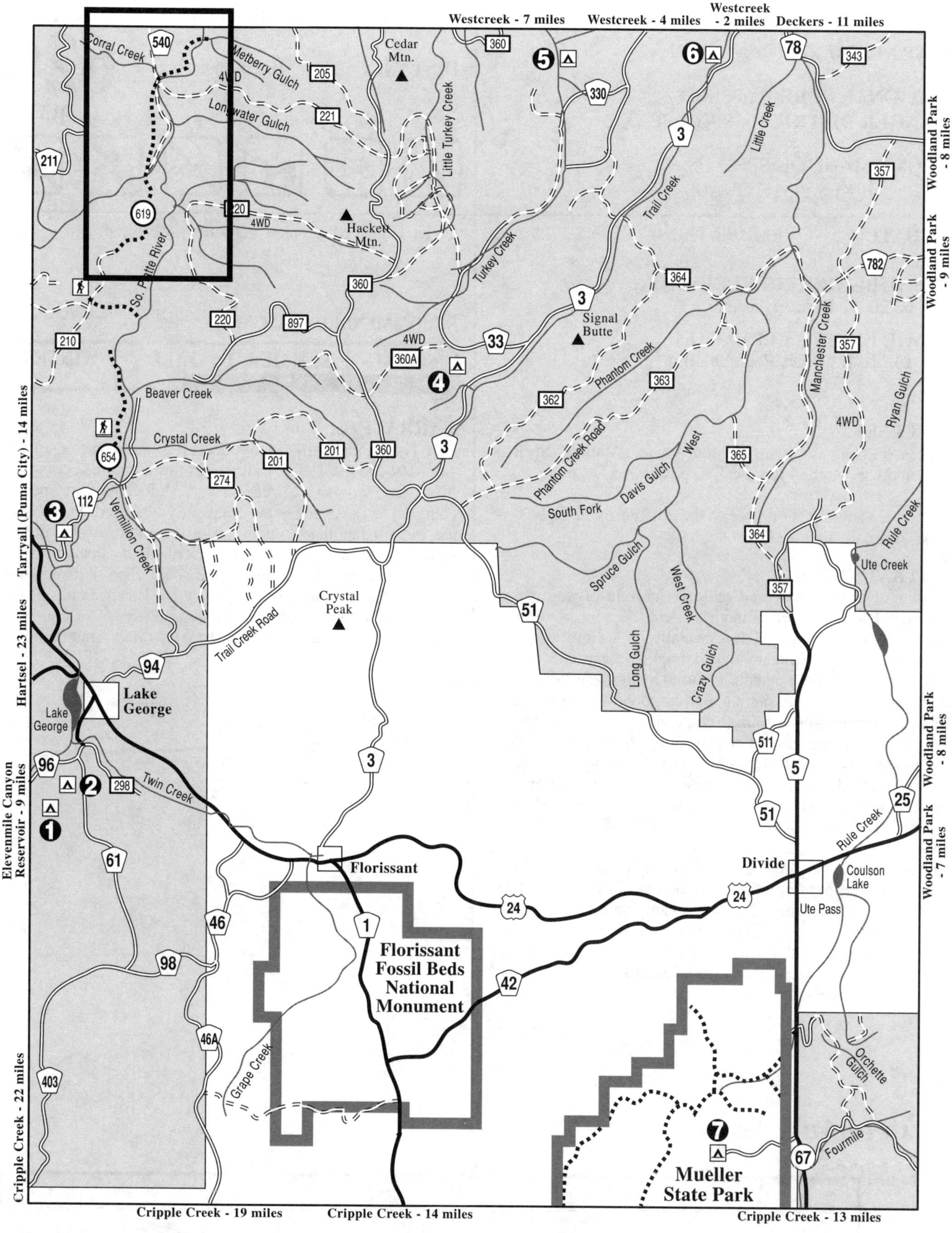

MAP 15

TRAIL #619

ROCKY MOUNTAIN REGION

MAP NUMBER(S): 19

NATIONAL FOREST: PIKE
RANGER DISTRICT: PIKES PEAK

USGS: MANITOU SPRINGS,
PIKES PEAK QUADS

DIFFICULTY: MORE DIFFICULT

DISTANCE: 13.0 MILES

TRAIL
620
BARR

UPDATED: NOVEMBER 1991

USE: Light, Motorized vehicles prohibited

TRAIL BEGINNING ELEVATION:
6,720 ft., Ruxton Avenue

TRAIL ENDING ELEVATION:
14,110 ft., Pikes Peak Summit

RECOMMENDED SEASON:

SPRING	SUMMER	FALL	WINTER

ACCESS:

Access 1. Trailhead is off Ruxton Avenue. Manitou Springs Colorado above the Pikes Peak Cog Railroad Depot.

Access 2. Some visitors ride to the summit by auto, bus, or Cog Railroad and hike down the trail.

ATTRACTIONS:

Hiking or riding along the Cog Railroad right-of-way is prohibited. The trail is rated more difficult because of its long sustained grade rising to an especially high elevation. Although an average hiker might complete the round trip in about 16 hours, those persons not used to hiking, especially at these altitudes, should not attempt the round trip in one day. Because of the long sustained grade, even a one way downhill hike from the summit is very tiring to the legs and feet when not conditioned. Temperatures at the summit may be as much as 40 degrees cooler than at the bottom. Afternoon thunderstorms are common throughout the summer and snow or sleet may be encountered throughout the year.

An open shelter is located about nine miles up the trail just below timberline. Campfires may be built at this site. Barr Camp is another popular overnight camping or lodging spot operated under a special use permit with the US Forest Service. A charge is made for use of the facilities. Drinking water is not readily available. Water from streams should be treated before use. Hikers are advised to carry an adequate supply.

NARRATIVE:

Barr Trail, named after its builder, Fred Barr was constructed from 1914 to 1921. The trail passes through three ecological life zones, Montane, Spruce-fir and Alpine. Wildlife species common to those zones are frequently observed. Pikes Peak has been a landmark since before recorded history. It was named for Zebulon Pike who described the Peak in 1806 while exploring In the area to the south. The first recorded climb to the summit was in 1820 by Dr. Edwin James of the Stephen H. Long Expedition. At present, over 60,000 visitors per year are estimated to climb the peak on the Barr Trail.

TRAIL PROFILE (ONE-WAY):

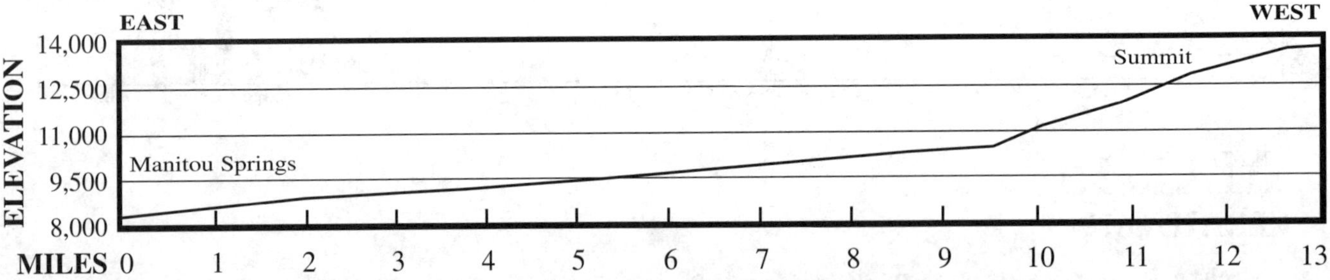

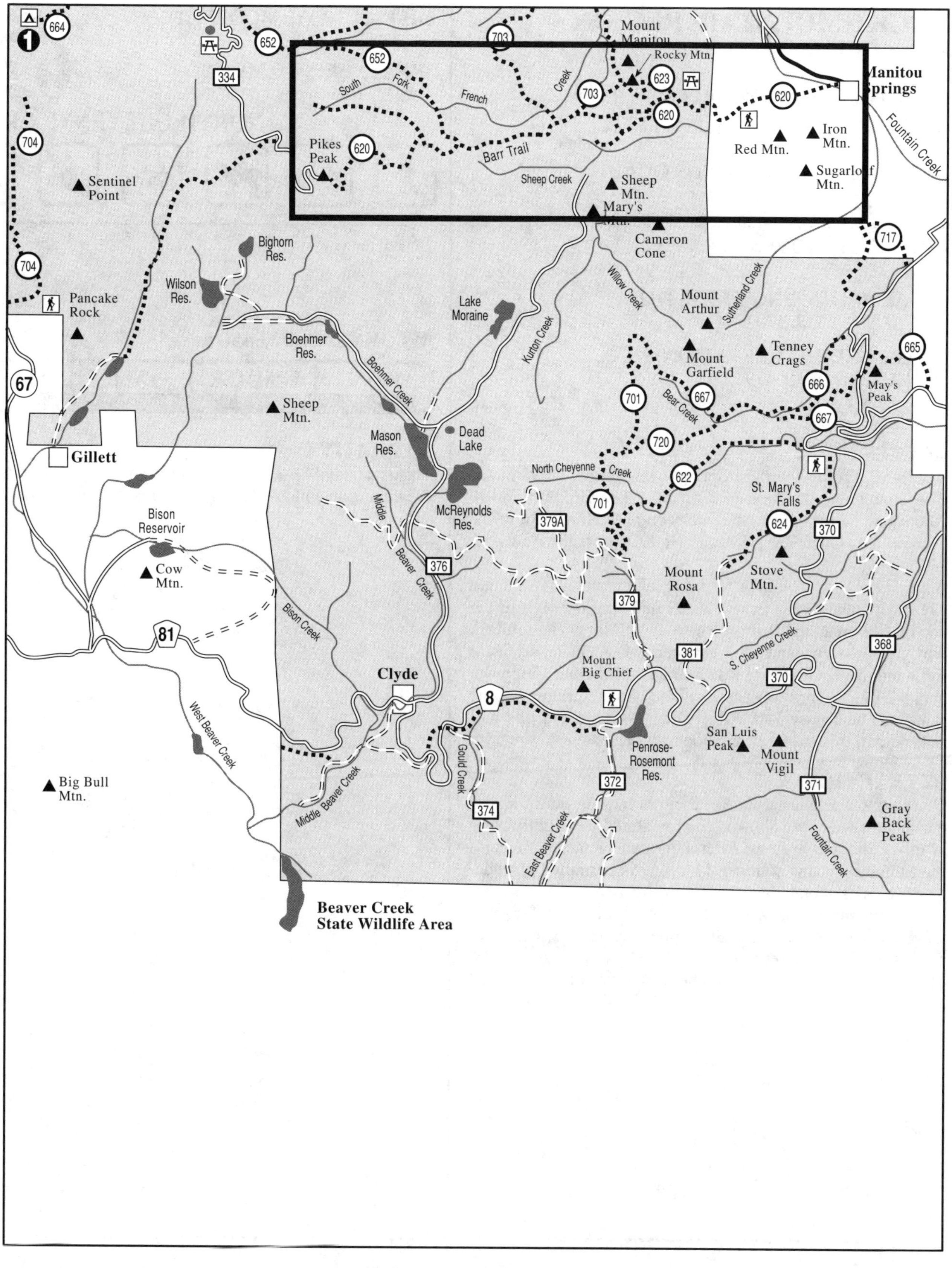

MAP 19
TRAIL #620

ROCKY MOUNTAIN REGION

MAP NUMBER(S): 19

NATIONAL FOREST: PIKE
RANGER DISTRICT: PIKES PEAK

USGS: MANITOU SPRINGS QUAD

DIFFICULTY: MODERATE

DISTANCE: 4.6 MILES

TRAIL 622
NORTH CHEYENNE CREEK

UPDATED: NOVEMBER 1991

USE: Heavy

TRAIL BEGINNING ELEVATION:
7,600 ft. FDR 370

TRAIL ENDING ELEVATION:
10,320 ft. FDR 379

RECOMMENDED SEASON:

SPRING	SUMMER	FALL	WINTER

ACCESS:
Access 1. From Colorado Springs, take either 26th St. or Cheyenne to North Cheyenne Canyon Rd. to Gold Camp Rd. About one half mile past the Intersection of North Cheyenne Canyon Rd.. Gold Camp Rd. and High Drive is the trailhead.

Access 2. From Colorado Springs, take either 21st St. and Cresta or Lake to the Broadmoor. Following the sign to the zoo on El Pomar and Penrose, the road will fork. The left fork will go to the zoo and the right fork is Old Stage Rd. Turn right and travel Old Stage Rd. until it ends at Gold Camp Rd. about 7 miles. Turn left and continue about 5 miles. On the right will be Frosty Park Rd. (FDR 379). One and one half miles down this road is where the trail begins.

ATTRACTIONS:
On the first one quarter mile of the trail leading out of Frosty Park motorcycles are allowed. Temperatures are slightly cooler than Colorado Springs. Afternoon thunderstorms are common throughout the summer. Camping is permitted at undeveloped sites along the trail. Campfire permits are not required but please be sure your fire is dead out before you leave. Pack it in/Pack it out rule applies. Drinking water is not readily available. Hikers are advised to carry an adequate supply. Water from streams should be treated before use.

NARRATIVE:
This is a heavily used trail. There are numerous side trails but the main trail follows North Cheyenne Creek to Frosty Park.

TRAIL PROFILE (ONE-WAY):

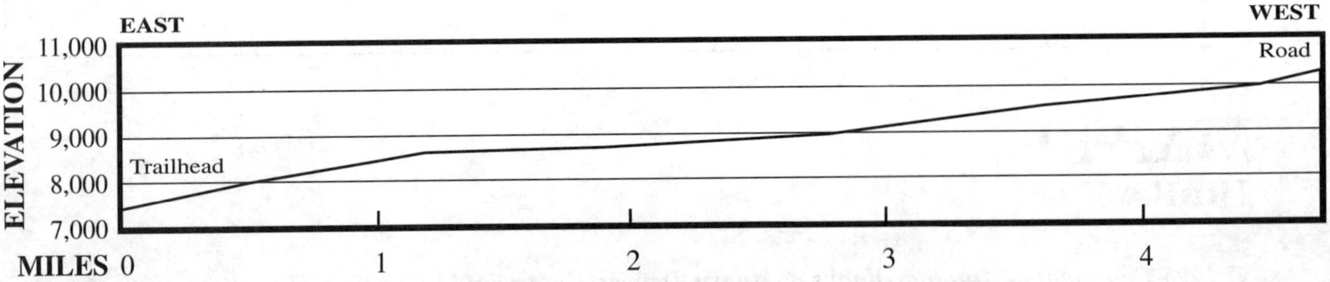

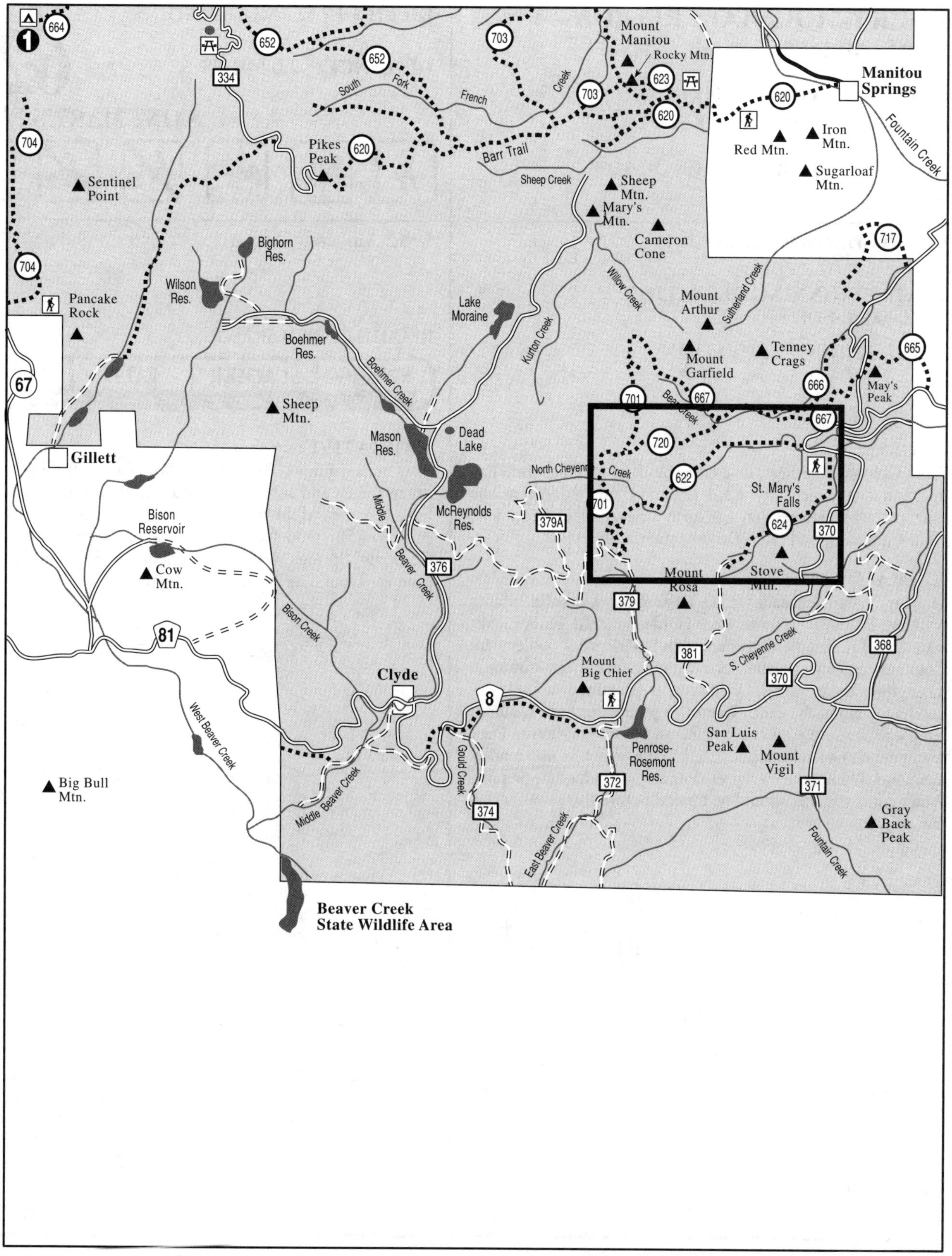

MAP 19

TRAIL #622

ROCKY MOUNTAIN REGION

MAP NUMBER(S): 19

NATIONAL FOREST: PIKE
RANGER DISTRICT: PIKES PEAK

USGS: MANITOU SPRINGS QUAD

DIFFICULTY: MODERATE

DISTANCE: 2.0 MILES

TRAIL 624

SAINT MARY'S FALLS

UPDATED: NOVEMBER 1991

TRAIL BEGINNING ELEVATION:
7,900 ft. FDR 370

TRAIL ENDING ELEVATION:
9,750 ft.

ACCESS:
From Colorado Springs, take either 26th St. or Cheyenne Rd. to North Cheyenne Canyon Rd. to Gold Camp Rd. About one mile past the Intersection to North Cheyenne Canyon Rd., Gold Camp Rd. and High Drive is the trailhead.

ATTRACTIONS:
St. Mary's Falls cascades 25,OOO feet down a solid granite wall and is about 2.0 miles from Gold Camp Rd. Falls Elevation: 8,880 ft. Temperatures are about 10 degrees cooler than Colorado Springs. Afternoon thunderstorms are common throughout the summer. Camping is permitted at undeveloped sites along the trail. Campfire permits are not required but please be sure your fire is dead out before you leave. Pack it in/Pack it out rule applies. Drinking water is not readily available. Hikers are advised to carry an adequate supply. Water from streams should be treated before use.

USE: Moderate, Motorized vehicles prohibited

RECOMMENDED SEASON:

SPRING	SUMMER	FALL	WINTER

NARRATIVE:
This trail follows Buffalo Canyon Creek through mature aspen stands and mixed pine and spruce forests to St. Mary's Falls. To one's right are the falls which cascade down a granite wall 250 - 300 feet. To one's left is a very nice view of Colorado Springs with the Broadmoor Hotel in the foreground. Trout may be seen in the creek.

TRAIL PROFILE (ONE-WAY):

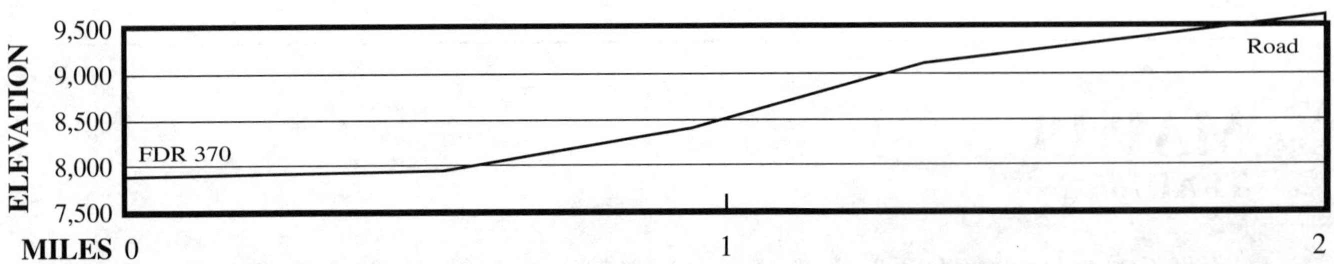

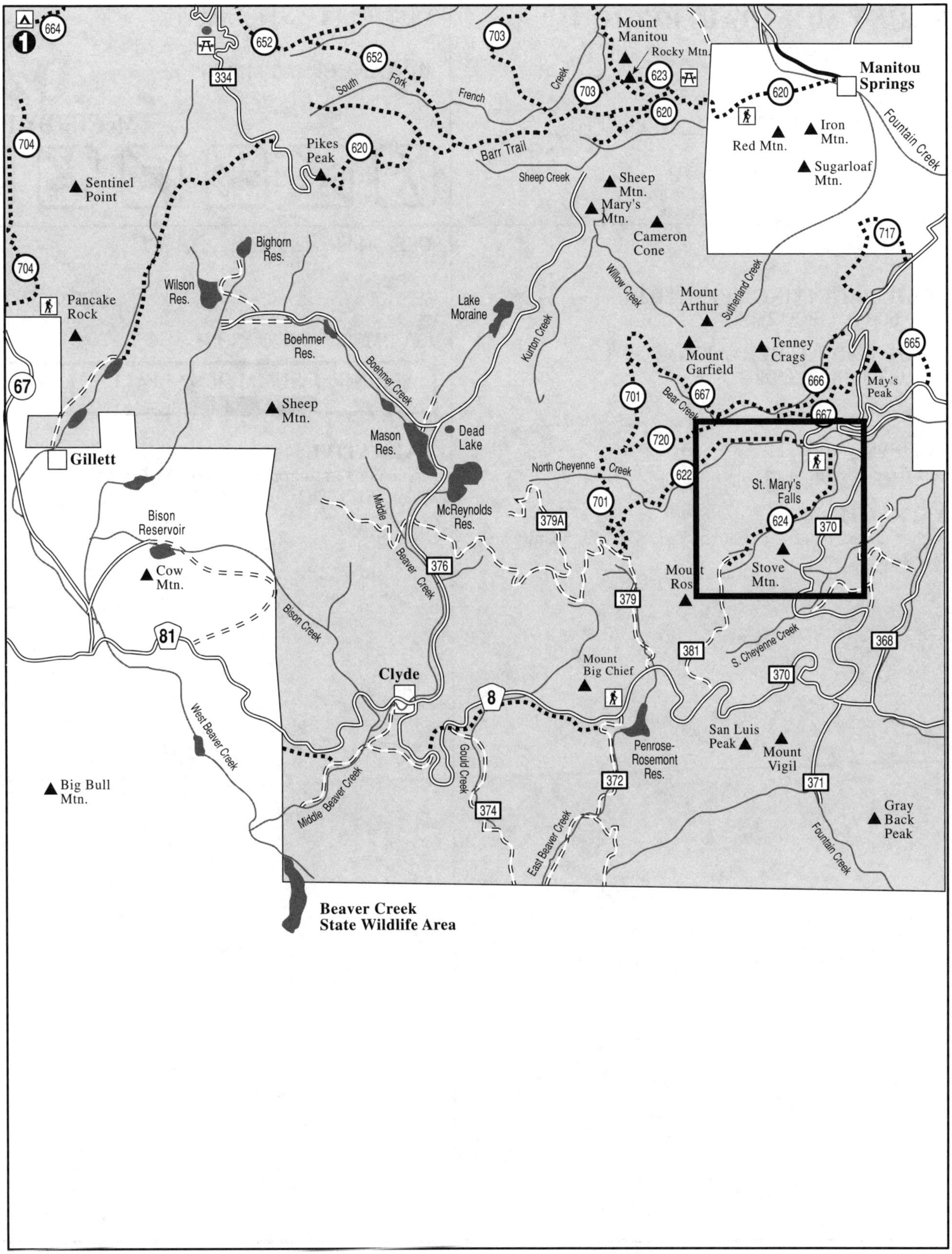

MAP 19
TRAIL #624

ROCKY MOUNTAIN REGION

MAP NUMBER(S): 9

NATIONAL FOREST: PIKE
RANGER DISTRICT: SOUTH PLATTE

USGS: WINDY PARK QUAD

DIFFICULTY: MODERATE

DISTANCE: 7.5 MILES

TRAIL 628
McCURDY PARK

UPDATED:

TRAIL BEGINNING ELEVATION:
9,600 ft. (Tr. 612)

TRAIL ENDING ELEVATION:
10,800 ft. (Tr.607)

ACCESS:
Lost Creek Scenic Area

ATTRACTIONS:
Shelter house, parking at trailheads. Lost Creek Scenic Area, back country ranger.

USE: Moderate to Heavy

RECOMMENDED SEASON:

SPRING	SUMMER	FALL	WINTER
	■		

NARRATIVE:
Connecting trail for Brookside McCurdy (607) and Goose Creek (612) Trails

TRAIL PROFILE (ONE-WAY):

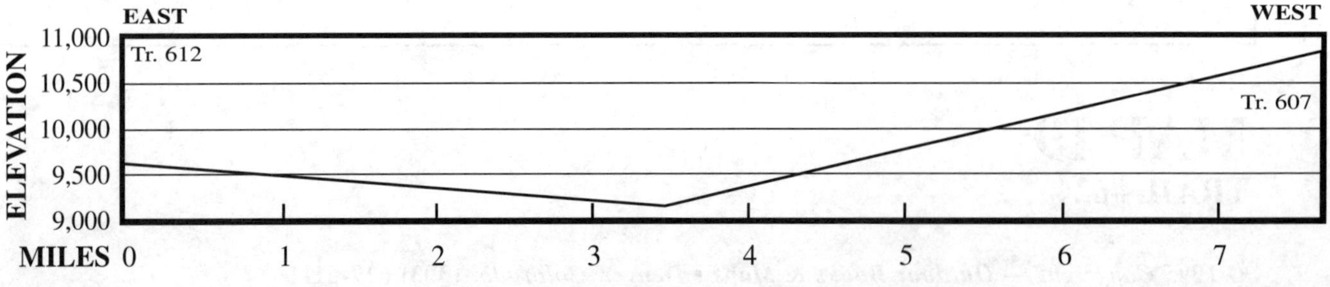

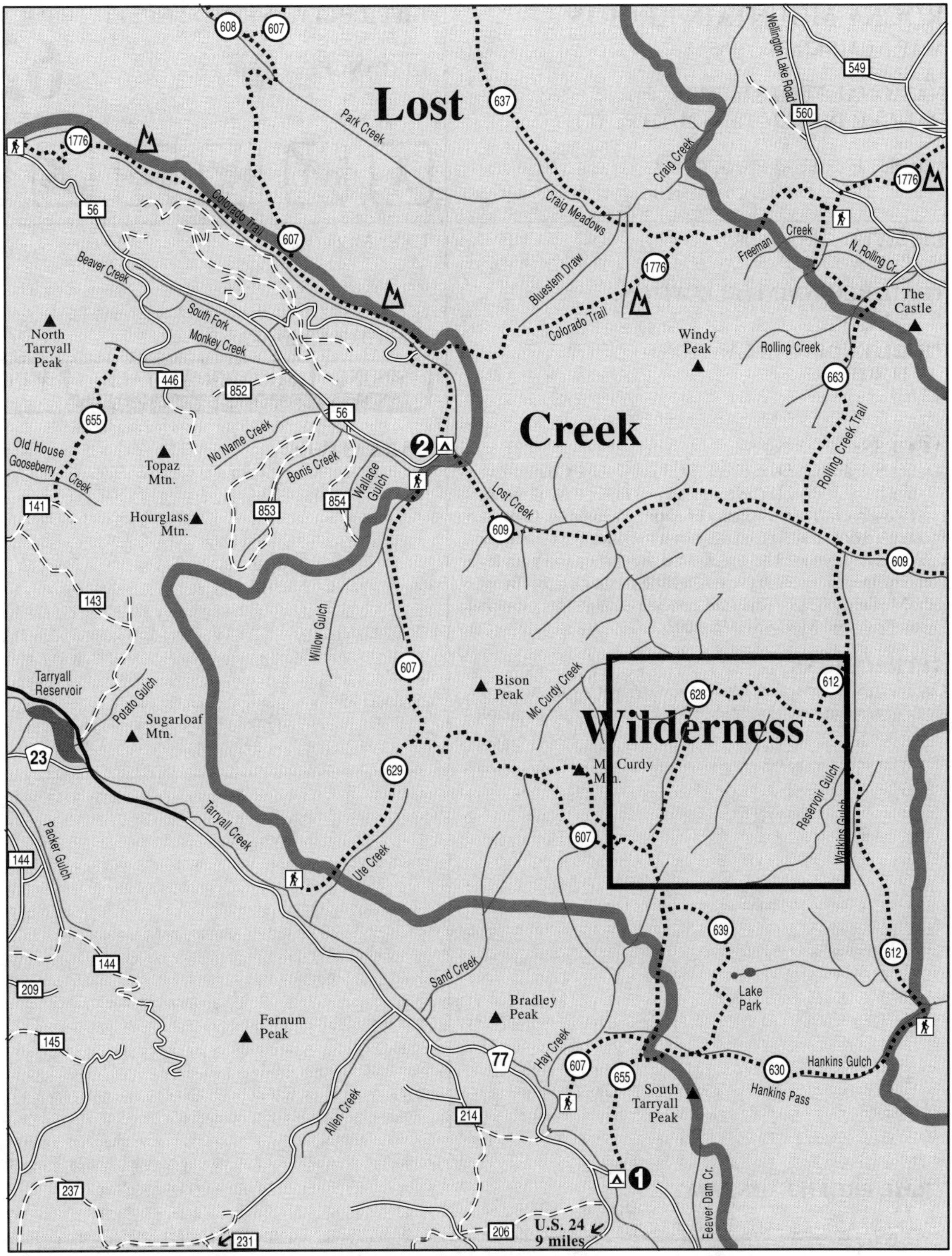

MAP 9

TRAIL #628

ROCKY MOUNTAIN REGION
MAP NUMBER(S): 9

NATIONAL FOREST: PIKE
RANGER DISTRICT: SOUTH PLATTE

USGS: FARNUM PEAK QUAD

DIFFICULTY: VERY DIFFICULT

DISTANCE: 4.0 MILES

TRAIL 629
UTE CREEK

UPDATED: MAY 1989

TRAIL BEGINNING ELEVATION:
8,300 ft.

TRAIL ENDING ELEVATION:
11,400 ft.

ACCESS:
Access by car to the Ute Creek Trail is off Park County Road 77, the Tarryall Road. The trailhead is 3 miles east of the Tarryall Reservoir. There is plenty of parking available. Cross the footbridge onto trail. From this point to Bison Pass it is a continuous steep grade. Ute Creek Trail provides a short route to some of the high country vistas a little farther up the Brookside McCurdy Trail. This trail provides access for climbing Bison Peak and McCurdy Mountain.

ATTRACTIONS:
On a warm summer day take extra water with you due to the steep grade and length of trail. Water is not readily available.

USE: Moderate

RECOMMENDED SEASON:

SPRING	SUMMER	FALL	WINTER

NARRATIVE:

TRAIL PROFILE (ONE-WAY):

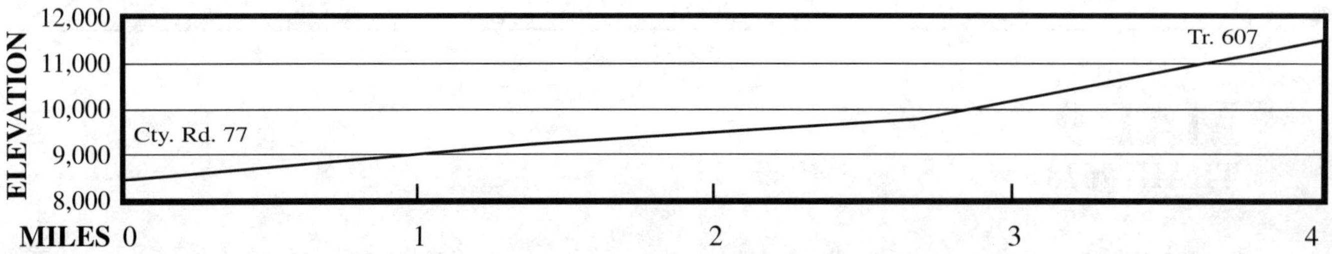

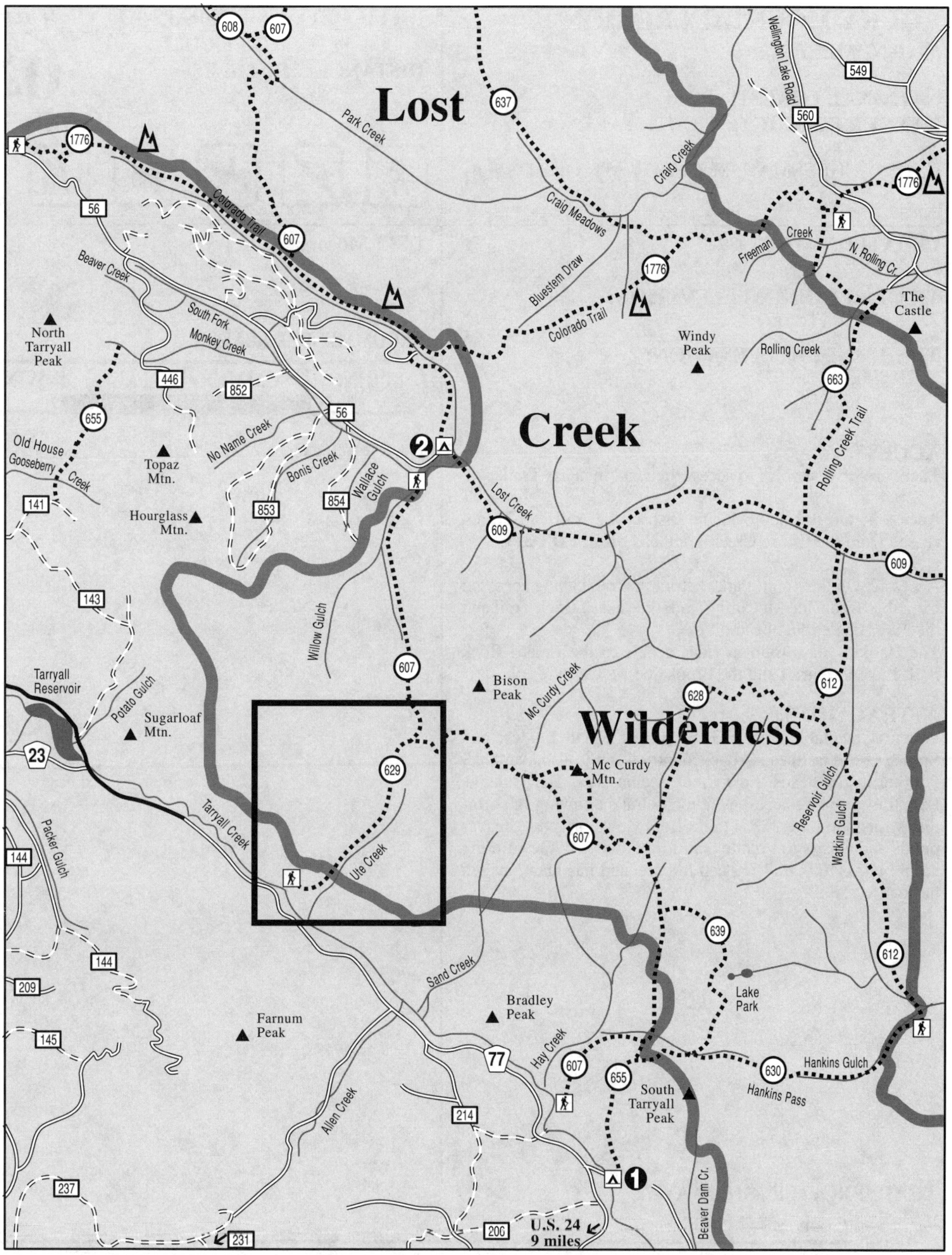

MAP 9

TRAIL #629

ROCKY MOUNTAIN REGION

MAP NUMBER(S): 9

NATIONAL FOREST: PIKE
RANGER DISTRICT: SOUTH PARK

USGS: CHEESMAN, McCURDY MT. QUADS

DIFFICULTY: MODERATE TO DIFFICULT
DISTANCE: 5.0 MILES

TRAIL 630
HANKINS PASS

UPDATED: MAY 1989

TRAIL BEGINNING ELEVATION:
8,200 ft. (Tr.612)

TRAIL ENDING ELEVATION:
8,800 ft. (Tr.607)

ACCESS:
There are several ways to access the Hankins Pass Trail.

Access 1. On the South Platte District the trail can be accessed from the Goose Creek Trailhead off of FDR 211.

Access 2. On the South Park District the trail can be accessed from the Brookside McCurdy Trail, the Lizard Rock Trail and the Twin Eagles Trailhead.
The Hankins Pass trail provides access to the Lizard Rock trail, Lake Park trail and the Brookside-McCurdy Trail.

ATTRACTIONS:
Water is available at the Goose Creek, Twin Eagles, and Spruce Grove trailheads. Backcountry water is also available. Parking at trailheads. Lost Creek scenic area. The Hankins Pass trail provides access with many trails allowing opportunities for loop hikes. This trail offers views of active beaver ponds and access to granite crag formations. While offering scenic vistas, this trail is steep in parts and has many switch backs.

USE: Moderate to heavy

RECOMMENDED SEASON:

SPRING	SUMMER	FALL	WINTER

NARRATIVE:

TRAIL PROFILE (ONE-WAY):

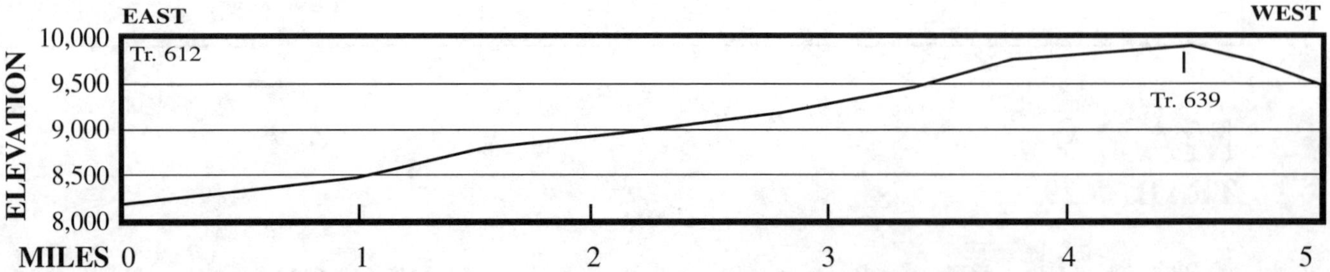

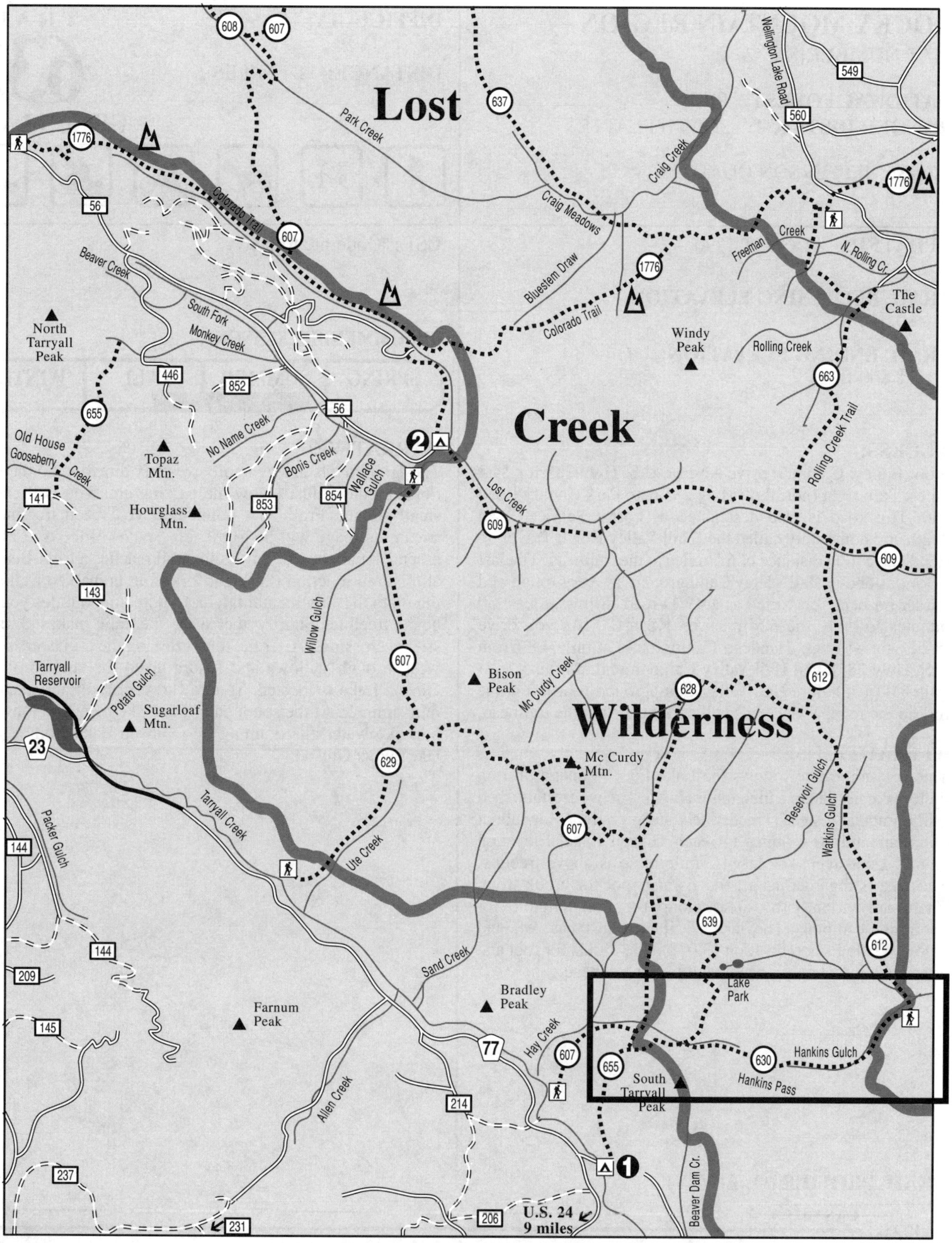

MAP 9
TRAIL #630

ROCKY MOUNTAIN REGION

MAP NUMBER(S): 2

NATIONAL FOREST: PIKE
RANGER DISTRICT: SOUTH PLATTE

USGS: JEFFERSON QUAD

DIFFICULTY: EASY

DISTANCE: 2.4 MILES

TRAIL 633
GIBSON LAKE

UPDATED: AUGUST 1990

TRAIL BEGINNING ELEVATION:
10,316 ft.

TRAIL ENDING ELEVATION:
11,860 ft.

USE: Moderate to heavy

RECOMMENDED SEASON:

SPRING	SUMMER	FALL	WINTER

ACCESS:
From Bailey, Colorado drive west on U.S. Hwy 285 for 14.3 miles. Turn right (north) off Hwy 285 onto Park County Road #60. This road is also designated as Forest Service Road #120. and commonly called the 'Hall Valley Road. Travel on F.S. #120 a total distance of 6.5 miles to the trailhead. The last 1.4 miles above Hall Valley Campground are very rough and border on being classified as a 4WD road. Allow at least 50 minutes to drive the 6.5 miles on F.S. #120. As you drive #120, you will pass Handcart Campground at mile 4.8 (from U.S. Hwy 285). and Hall Valley Campground at mile 5.1. At mile 5.0 (just before Hall Valley Campground bear left at the fork in the road. There is a small parking area at the trailhead.

ATTRACTIONS:
This 2.4 -mile hike follows the Lake Fork drainage up to a pretty lake nestled in a high tundra bowl. The well-established trail climbs at a steady, moderately steep grade. Allow about one hour and 20 minutes to reach Gibson Lake. Elevation gain is 1.544 feet. The last 1/2 mile or so is above treeline. The lake is regarded as a good fishing spot for brook trout. There are virtually no areas along the trail suitable for overnight camping. The lake is situated just east of, and below Whale Peak (elevation 13.078). The trail is for foot and horseback travel only. Picnic facilities at trailhead.

NARRATIVE:
The Trail travels in a generally westerly direction up the Lake Fork drainage. Just below the parking area you will cross a small stream. From that point until you reach treeline the stream drainage will be to the left (south) of the trail as you ascend. At about 1 3/4 miles the trail reaches a fork. Bear left at this trail junction (the right fork is an unmarked trail leading up to the Continental Divide). At about 2.0 miles you will be at treeline. Shortly thereafter the trail makes 3 major stream crossings. After the third crossing the trail bears southwest through willows and tundra up to the shelf on which Gibson Lake is located. Tundra flowers are in abundance in mid-summer. At the south end of the lake there is a long, ribbon-like waterfalls draining into Gibson Lake from a small lake above Gibson.

TRAIL PROFILE (ONE-WAY):

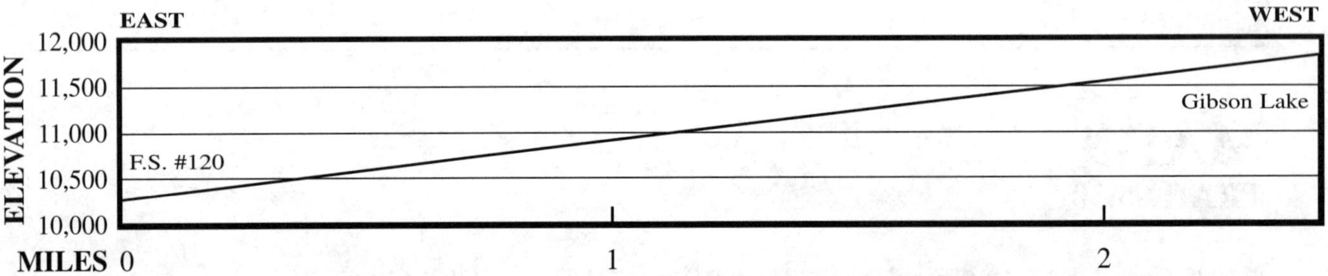

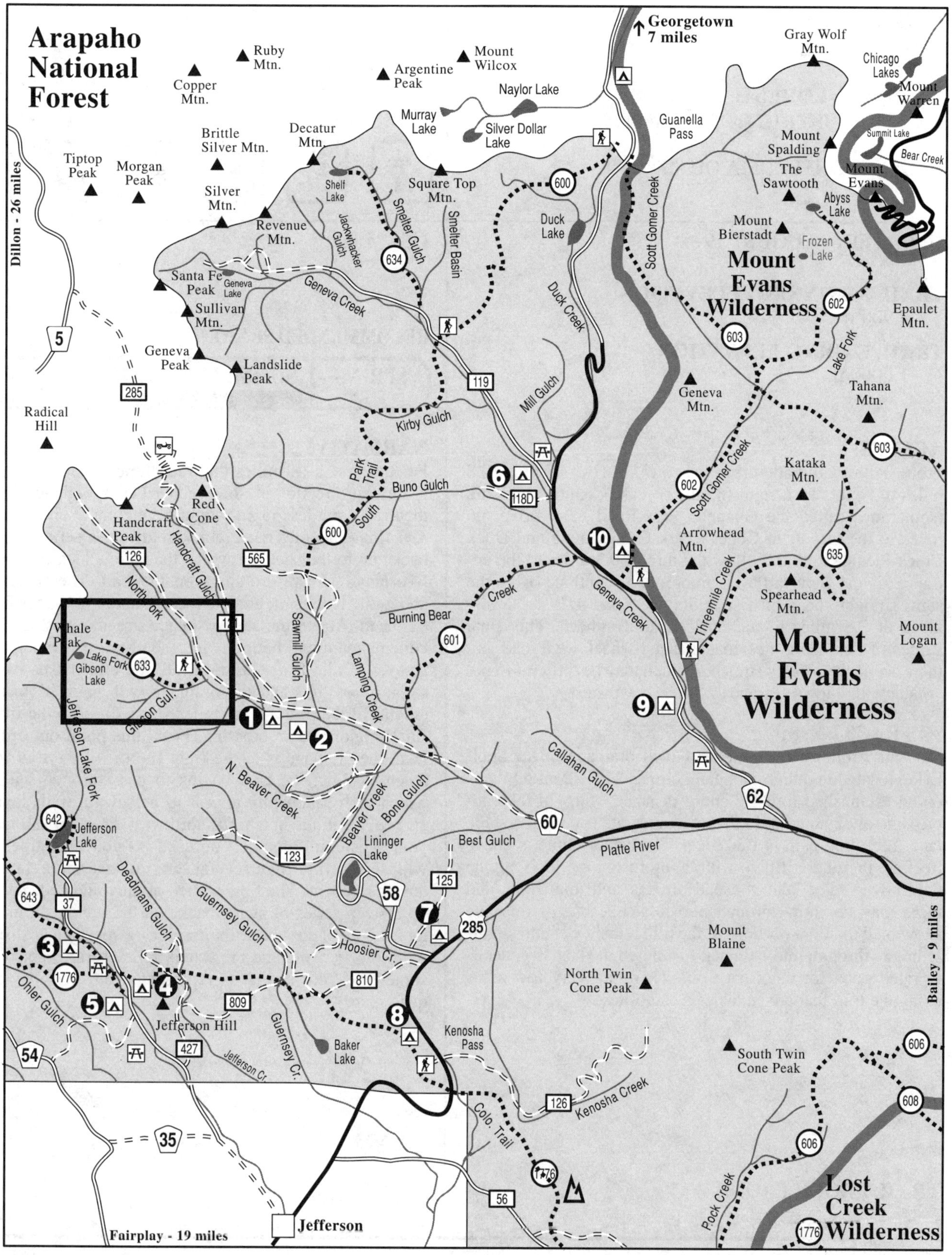

MAP 2

TRAIL #633

ROCKY MOUNTAIN REGION

MAP NUMBER(S): 2

NATIONAL FOREST: PIKE
RANGER DISTRICT: SOUTH PLATTE

USGS: MONTEZUMA QUAD

DIFFICULTY: EASY

DISTANCE: 3.2 MILES

TRAIL 634
SHELF LAKE

UPDATED: AUGUST 1990

TRAIL BEGINNING ELEVATION:
9,900 ft.

TRAIL ENDING ELEVATION:
12,000 ft.

USE: Light

RECOMMENDED SEASON:

SPRING	SUMMER	FALL	WINTER

ACCESS:
From Bailey, Colorado drive west on U.S. Hwy 285 for 10.8 miles to Grant. Turn right (north) on Park County Road #62 (sometimes called the Guanella Pass Road), and drive 6.6 miles to the turnoff to Geneva Park Campground and Duck Creek Picnic Ground. Follow the turnoff 0.3 miles to the entrance sign at Geneva Park Campground. Continue (from the sign) straight ahead on Forest Service Road #119 for a distance of 2.8 miles to the Shelf Lake Trailhead. This final stretch of road is not designated as a 4-wheel drive road, but there are several rocky stretches which may be difficult to negotiate with a low-clearance 2-wheel drive auto.

ATTRACTIONS:
This hike follows the Smelter Gulch drainage up to Shelf Lake, located in a high mountain cirque. The 3.2 mile hike is on an established trail the entire distance, and will take approximately 2 hours of hiking at a moderate pace to reach the lake. Elevation gain is 1,842 feet. The last 3/4 mile is above treeline in tundra filled with a profusion of high alpine wildflowers. The lake is usually frozen until mid-June, and there may be snow through mid-July. Because of this late snow melt the final portions of the trail remain wet and slightly boggy through mid-summer. Fishing in the lake is generally rated good for cutthroat trout. There are very few areas along the trail that are suitable as campsites.

NARRATIVE:
For the first 1 1/2 miles the trail climbs steadily in a north-northwest direction in the Smelter Gulch drainage. The trail then rises at a less rapid rate until treeline at about mile 2.4. The final 3/4 mile rises more rapidly in a series of switchbacks up to the shelf on which the lake is located. There are five major stream crossings on the trail. The first two are about 200 feet apart, and are located about 0.7 miles from the trailhead. Above the second major stream crossing, Smelter Gulch stream will be to the left (south) of the trail. At slightly over 1 mile from the trailhead, the trail makes two more stream crossings and the drainage will again be to the left (south) of the trail. At 2.0 miles you will reach the fifth (and final) major stream crossing. From this point onward to the lake, the drainage is to the right (north) of the trail. As you reach treeline, the trail begins to pass through willows in which there frequently appear to be several trails that come and go. Continue in a north-northwest direction and maintain your elevation if you get confused. As you leave treeline you will see on your right several series of cascading, falls as the terrain steepens. The final stretch of the trail bends left (west) up a steep series of switchbacks on the tundra. In mid-summer this final stretch of the trail has a spectacular display of wildflowers. Note that the actual trail described above varies somewhat from how it is represented on the topographical quad map.

TRAIL PROFILE (ONE-WAY):

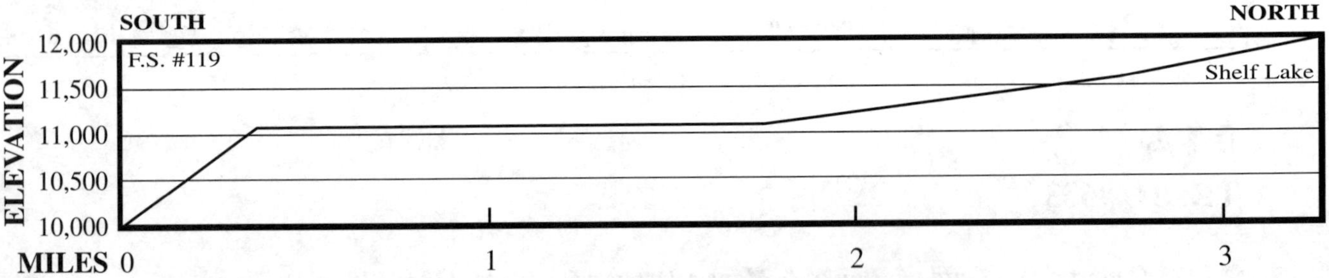

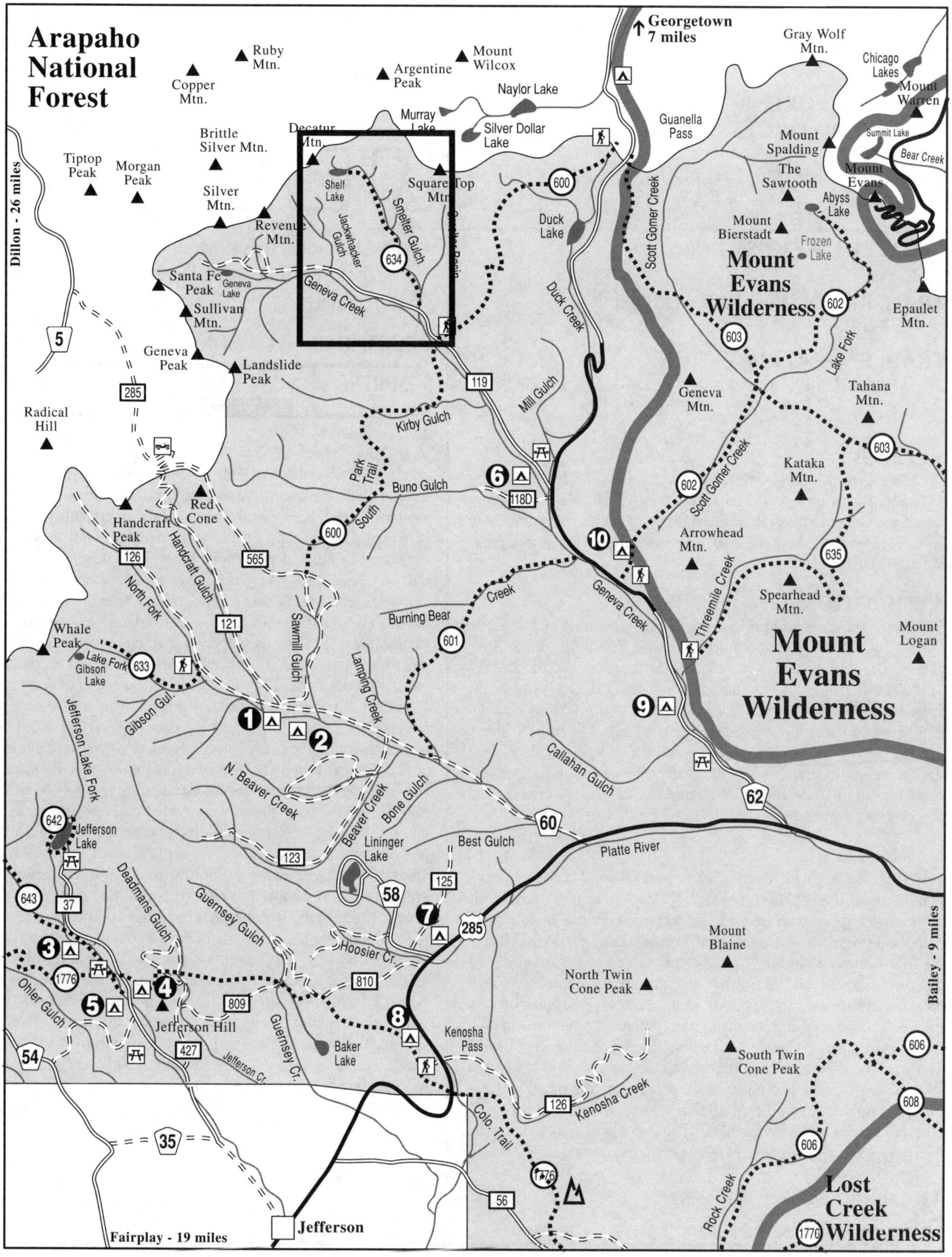

MAP 2

TRAIL #634

ROCKY MOUNTAIN REGION
MAP NUMBER(S): 2

NATIONAL FOREST: PIKE
RANGER DISTRICT: SOUTH PLATTE

USGS: MT. EVANS, MT. LOGAN QUADS

DIFFICULTY: MODERATELY DIFFICULT
DISTANCE: 6.5 MILES

TRAIL 635
THREE MILE CREEK

UPDATED: AUGUST 1990

TRAIL BEGINNING ELEVATION:
8,980 ft.

TRAIL ENDING ELEVATION:
11,640 ft. junction of Rosalie Trail

USE: Moderate/Heavy, Mountain bikes and other motorized/mechanical vehicles prohibited

RECOMMENDED SEASON:

SPRING	SUMMER	FALL	WINTER

ACCESS:
Drive west from Bailey, Colorado on U.S. Hwy 285 for 10.8 miles to Grant. Turn right (north) on Park County Road #62 (sometimes called the Guanella Pass Road), and drive 2.8 miles to the trailhead on the right (east) side of the road. There is a limited amount of parking at the trailhead.

ATTRACTIONS:
This popular hiking trail provides access into the heart of the southern half of the Mt. Evans Wilderness Area. The trailhead is along Geneva Creek Road and its terminus is the junction with the Rosalie Trail. The 6.5 mile hike is rated as moderately difficult due to its length and elevation gain of 2,840 feet. The trail offers spectacular high mountain scenery and vistas. The last mile of the trail is above treeline. Allow 3 1/2 - 4 hours to reach the junction with Rosalie Trail using a moderate, steady pace, and 3 hours to return. Because of its location In a designated Wilderness Area mountain bikes and other motorized/mechanical vehicles are prohibited.

NARRATIVE:
The Three Mile Creek Trail is not shown on the Mt. Evans topographic quad map (photo revised 1974 copy), so this narrative is written with specific reference to marker points on the quad map to help locate your position and that of the trail. The first three miles of the trail follows alongside Three Mile Creek in a general northeast direction. The stream and trail are in a narrow canyon, with the trail crossing the stream at a dozen or so log bridges. At approximately mile 3 (1 1/2 hours of hiking from the trailhead), you come upon a small open meadow with a rock cairn and signpost saying "Trail". The stream crossing just before entering the meadow will be the last water source for the remainder of the trail. From the meadow signpost bear right (easterly direction) into a grove of aspen. In about 15-20 minutes you will reach another "Trail" signpost. From this point the trail rises steeply in a series of switchbacks moving up the southwest flank of Peak 12574. This section is quite steep and will take about 25-30 minutes to climb. As you ascend on these switchbacks look behind you from time to time for dramatic views to the southwest of Kenosh Pass, South Park, and the mountains west and south of Fairplay. At the top of the switchbacks the trail begins heading northwest and then north around the west flanks of Peak 12574. You will still be in trees and gaining elevation, but at a much lower rate than the steep switch backs. After about 20 minutes of hiking from the last switchback you will emerge from the trees on the northwest flank of Peak 12574. As you emerge into open tundra you will note the following landmarks: Kataka Mountain will be on your left (west), Mt. Evans and Mt. Bierstadt will be ahead of you and slightly to the left (north-northwest), and below you is a bowl-shaped depression that opens downward to your left. You will also be at the highest point on the trail. From this point you need to get to the opposite side of the bowl (on a line looking toward Bierstadt), but if you proceed directly across the depression you will be in an interminable willow bog. For this reason the trail has been routed to stay up out of the bog by traversing the east and north ridges of the depression. Look for large rock cairns placed in the willows and on the tundra. When you reach the northwest edge of the depression (the grassy tundra is almost flat here) the trail then drops down for the final 1/3 mile to the Rosalie Trail junction, marked by a signpost. At the trail junction pause to look to the east down into Deer Creek Valley. This is the general direction the Rosalie Trail follows down to its trailhead near Deer Creek Campground (a distance of about 6 miles). If you follow Rosalie Trail to the west it intersects the Abyss Lake Trail in about 2 mile, and terminates at Guanella Pass after an additional 4 miles of hiking. If you plan to return to the Three Mile Creek Trailhead follow the same route you used coming up to the junction point.

TRAIL PROFILE (ONE-WAY):

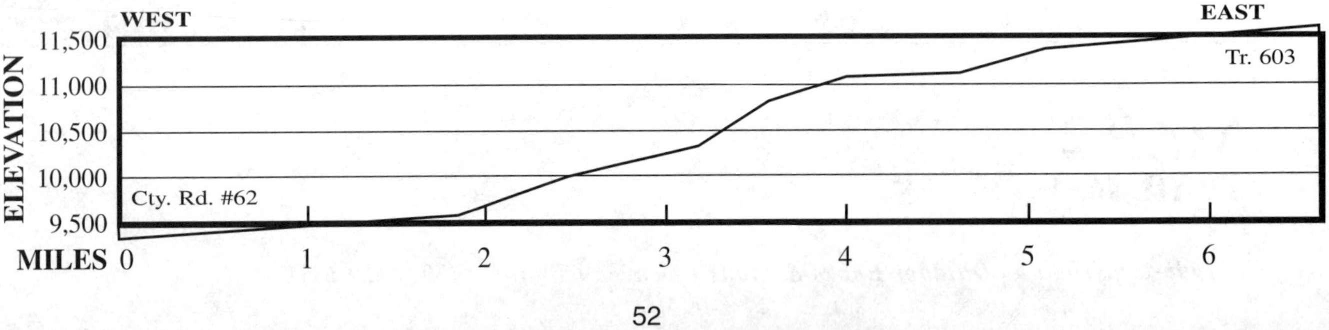

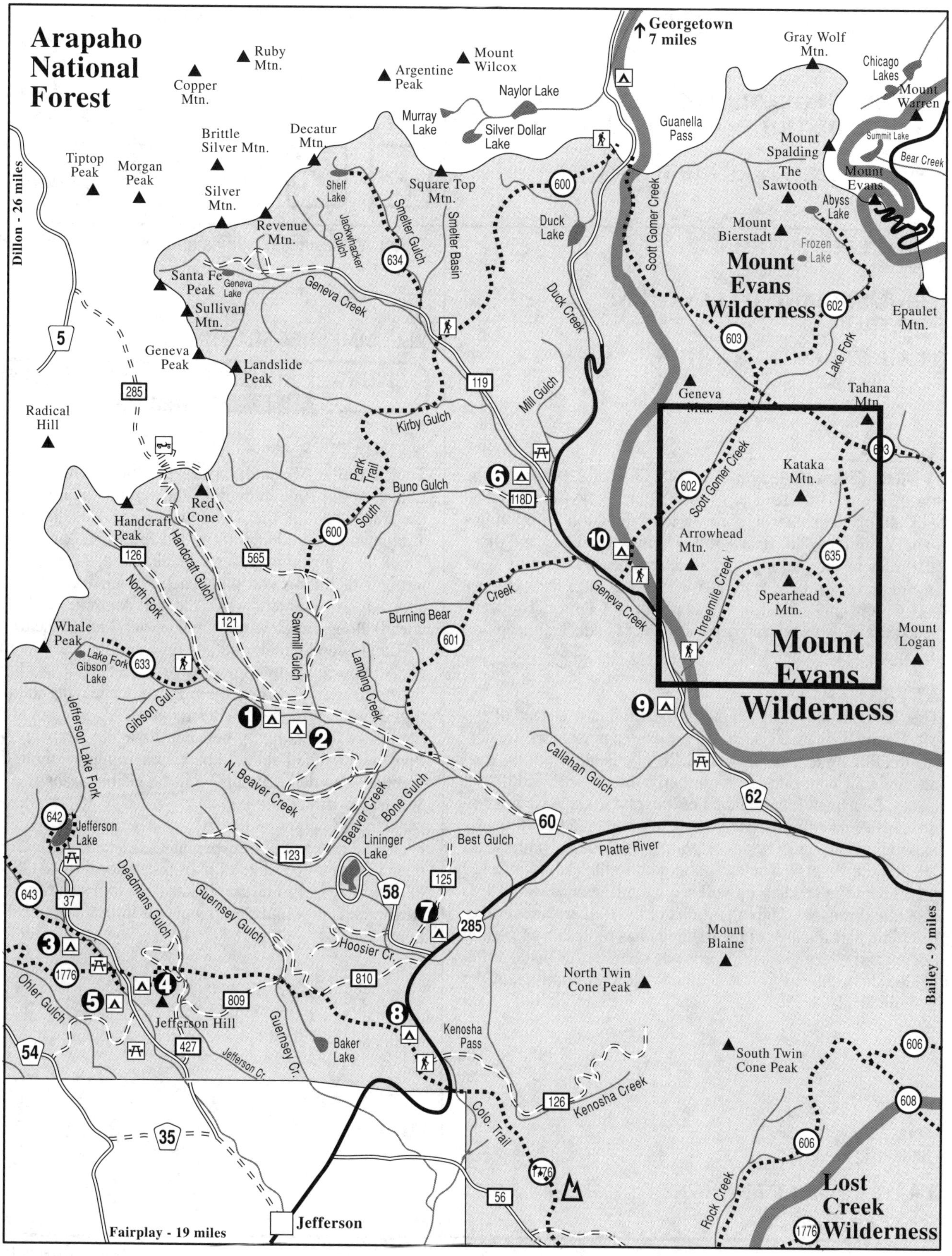

MAP 2
TRAIL #635

ROCKY MOUNTAIN REGION

MAP NUMBER(S): 3

NATIONAL FOREST: PIKE
RANGER DISTRICT: SOUTH PLATTE

USGS: HARRIS PARK QUAD

DIFFICULTY: MODERATE

DISTANCE: 4.0 MILES

TRAIL 636
TANGLEWOOD

UPDATED: SEPTEMBER 1990

TRAIL BEGINNING ELEVATION:
9,320 ft.

TRAIL ENDING ELEVATION:
11,742 ft.

USE: Moderate, Bicycles and vehicles prohibited

RECOMMENDED SEASON:

SPRING	SUMMER	FALL	WINTER

ACCESS:

Drive east from Bailey on U.S. Hwy 285 for 2.5 miles to the top of Crow Hill. Turn left (north) onto Park County Road #43 and drive in a north/ northwesterly direction for 6.8 miles to a "Y" in the road. Bear left on County Road #43 and drive 2.1 miles to the parking area at the trailhead. If driving west from Denver on U.S. Hwy 285 the turnoff onto Park County #43 is 10.8 miles west of the stoplight in Conifer. The trailhead parking area is designated as "Deer Creek Trailhead" on the signs along the road.

ATTRACTIONS:

The Tanglewood Trail begins in the southeast corner of the Mt. Evans Wilderness Area and heads in a north/northwesterly direction up to a saddle east of Rosalie Peak. From the saddle the trail continues in a northerly direction to Roosevelt Lakes. Beartrack Lakes and points beyond in the Arapaho National Forest. Immediately east of the saddle (towards Rosedale Peak) are a series of pointed rock outcroppings call "Pegmatite Points". The elevation gain to the saddle is 2,620 feet. From the saddle you will see magnificent views of two separate drainages. Upper portions of the trail are above treeline. The parking area at the trailhead has room for unloading horses and overnight camping is not permitted. This location is also the trailhead for the southeast (lower) terminus of the Rosalie Trail.

NARRATIVE:

The narrative below describes the trail as far as Roosevelt Lakes. About 100 feet west of the parking area you will reach the trailheads for the Tanglewood and Rosalie Trails. The Tanglewood Trail bears right and follows along Tanglewood Creek in a north/northwesterly direction. After about 25-30 minutes of hiking you will reach an unmarked fork. An old access road bears left, while the trail continues right (straight ahead) along Tanglewood Creek. After 3 miles the trail crosses Tanglewood Creek one last time and then begins to steepen. As you reach treeline you will see two of the Pegmatite Points straight ahead on the ridge. After leaving the trees the trail is quite steep and fairly strenuous. A post cairn at the saddle marks the boundary between Pike and Arapaho National Forests. Rosalie Peak will be visible immediately to the west and the "Pegmatite Points" will be visible along the ridge in an easterly direction.

From the saddle the trail begins to descend gradually across open tundra to Roosevelt Lakes (elevation 11.742), a distance of 0.4 miles. From the Lakes the trail continues to the Beartrack Lakes and points beyond in the Arapaho National Forest.

TRAIL PROFILE (ONE-WAY):

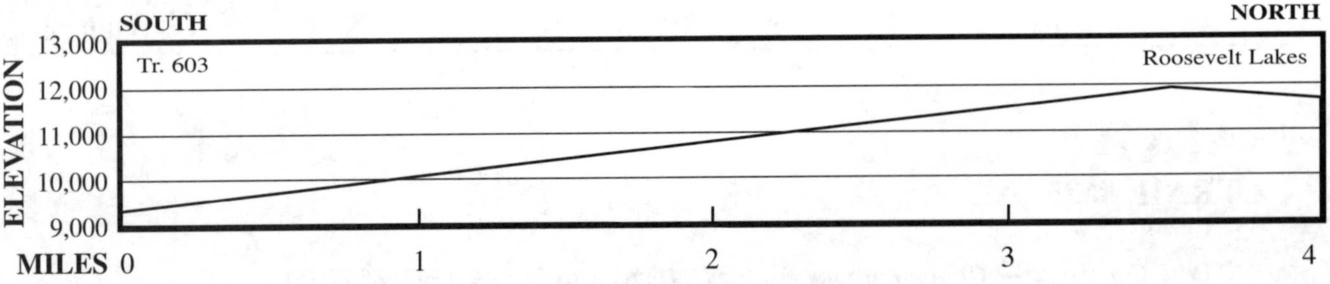

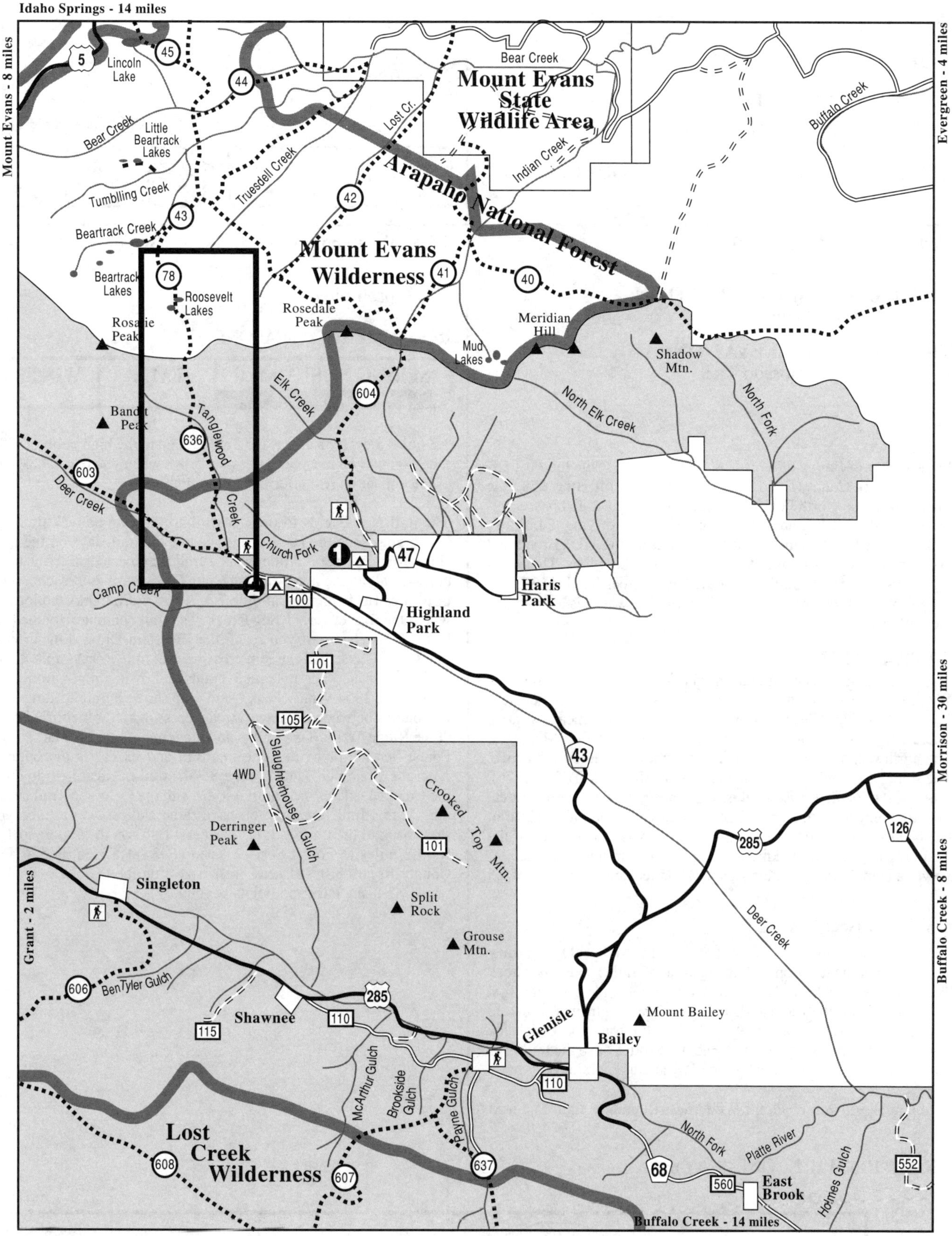

MAP 3

TRAIL #636

ROCKY MOUNTAIN REGION
MAP NUMBER(S): 3 and 9

NATIONAL FOREST: PIKE
RANGER DISTRICT: SOUTH PLATTE

USGS: SHAWNEE, WINDY PEAK QUADS

DIFFICULTY: MODERATE TO DIFFICULT
DISTANCE: 6.5 MILES

TRAIL 637
CRAIG MEADOWS

UPDATED: APRIL 1992

TRAIL BEGINNING ELEVATION:
8,240 ft. at Trailhead

TRAIL ENDING ELEVATION:
9,315 ft. at Colorado Trail

USE: Foot and horse travel only. Bicycles and vehicles prohibited.

RECOMMENDED SEASON:

SPRING	SUMMER	FALL	WINTER

ACCESS:
In the center of Bailey turn off U.S. Hwy #285 opposite the Post Office onto Park County Road No. 64. As you turn off Hwy. #285 the road crosses the North Fork of the South Platte River and passes between the Post Office and Moore Lumber. After driving 0.45 miles you will reach a "Y" in the road. Bear left on the gravel road and drive an additional 1.6 miles to the trailhead parking area. This parking area serves as the trailhead for the Payne Creek Trail (#637) and the Brookside McCurdy Trail (#607). Drinking water is available at the trailhead.

ATTRACTIONS:
This trail provides access into the northeast portion of the Lost Creek Wilderness Area. A highlight of the hike is a lovely area called Craig Meadows. Craig Creek passes through this open meadow area. There are sites suitable for overnight camping in Craig Meadows. The northwest terminus of the trail is the trailhead described above. The southeast terminus is a junction with the Colorado Trail (#1776). Most of the Payne Creek Trail is in the designated wilderness area, therefore bicycles and all motorized vehicles are prohibited. The trail has been identified on some maps as the Payne Gulch Trail or the Craig Meadows Trail. Hiking the entire trail from north to south involves a cumulative elevation gain of 2,285 feet and a distance of 9.5 miles.

NARRATIVE:
This trail description is based on starting at the trailhead near Bailey and ending at the junction with the Colorado Trail. The Payne Creek Trail and the Brookside McCurdy Trail are coincident for the first 1.75 miles from the trailhaed. Portions of this segment follow old roads. At mile 1.75 bear left at the fork in the trail (the Brookside McCurdy Trail bears right). For the next 1.5 miles or so the trail rises gradually along the west flank of the Payne Gulch drainage.

At mile 3.4 you will reach the Wilderness Boundary sign. The next 1.3 miles are very steep as the trail rises up to the high point of 9,890 feet. You will be in a pleasant aspen grove when you reach 9,890 feet - a nice place to rest after the steep climb!

The trail now begins to drop down into the Craig Creek drainage. When you reach the large open meadow the trail stays on the north side of Craig Creek. Although the trail may be indistinct at times, continue along the north side og Craig Creek until you reach a series of post cairns beginning at mile 7.3. A post cairn marks the location where the trail crosses Craig Creek. The trail continues southeasterly for 0.5 miles until you reach the Bluestem Draw flowing north into Craig Creek. This stream crossing is at mile 8.0. At mile 8.2 the trail crosses another unnamed drainage flowing north into Craig Creek. The 1954 Windy Peak Quad Map shows a trail heading south that begins between these two stream crossings. It is identified as the "Pine Ridge Trail". This trail is no longer maintained or marked. The Forest Service plans to close this trail and its use is discouraged. After the stream crossing at mile 8.2 the trail begins to climb out of the Craig Meadows area until it joins with the Colorado Trail at mile 6.5. At this trail junction you have three choices: (1) retrace your steps back to the trailhead; (2) head east (left) on the Colorado Trail for 3.2 miles to Forest Service Road #543 (also identified as Park County Road #88); (3) head west (right) on the Colorado Trail towards Kenosha Pass and points beyond.

TRAIL PROFILE (ONE-WAY):

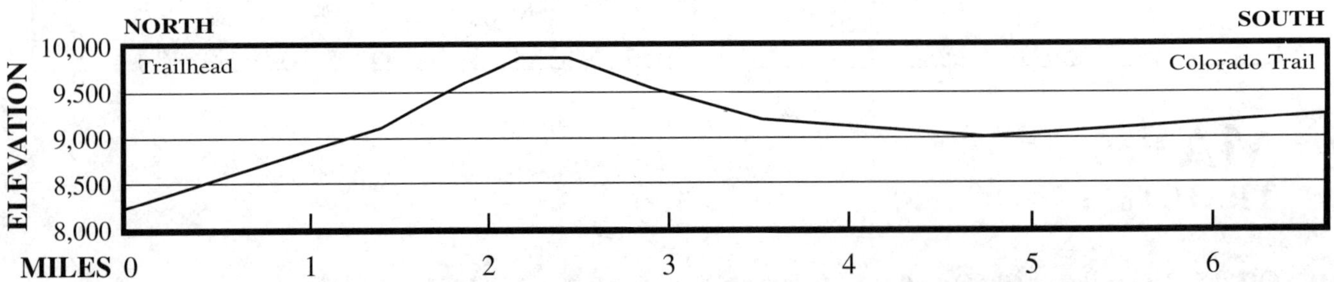

MAP 3 - South Half

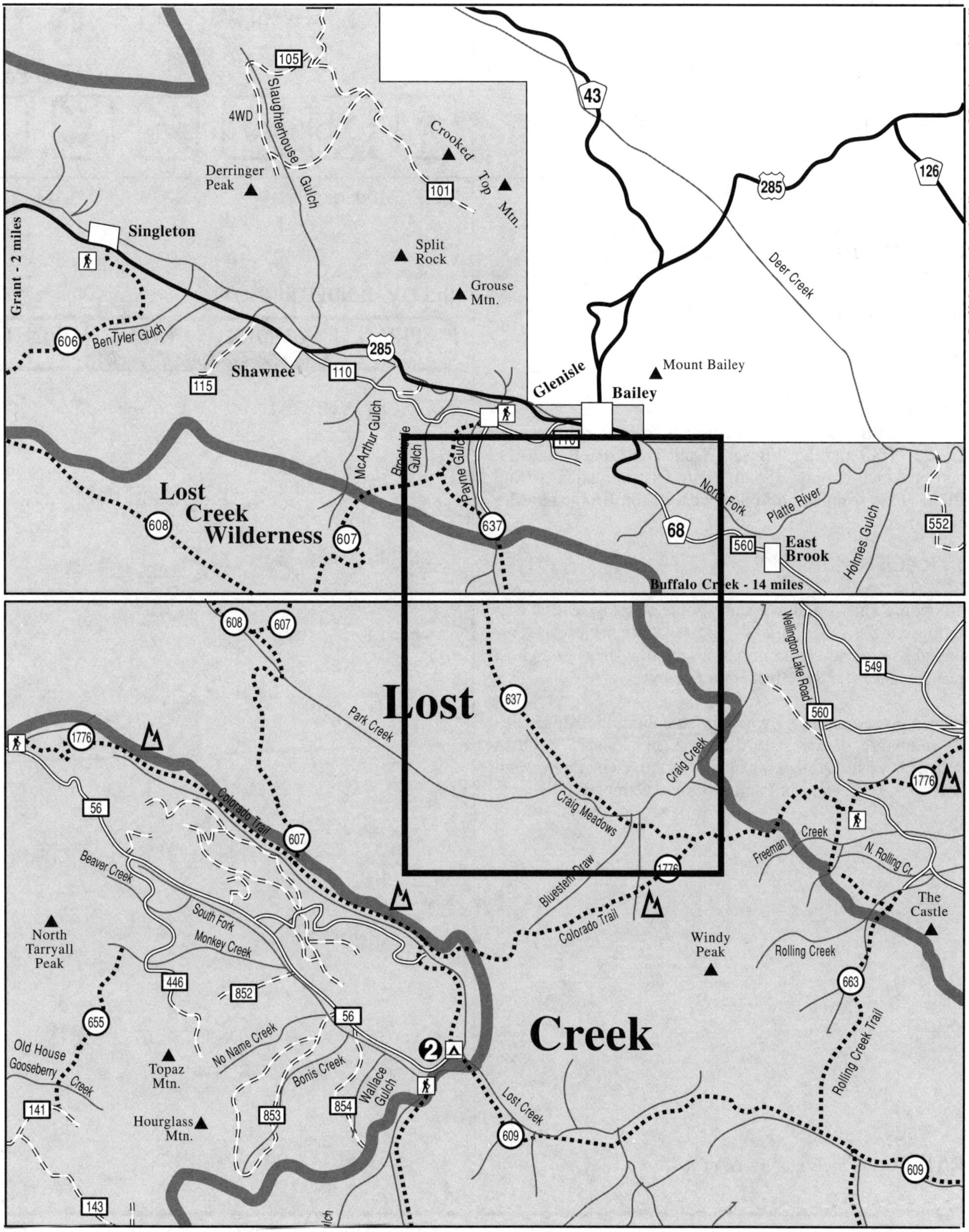

MAP 9 - North Half

TRAIL #637

ROCKY MOUNTAIN REGION

MAP NUMBER(S): 9

NATIONAL FOREST: PIKE
RANGER DISTRICT: SOUTH PARK

USGS: McCURDY MTN., WINDY PEAK QUADS

DIFFICULTY: MODERATE TO DIFFICULT
DISTANCE: 4.0 MILES

TRAIL
639
LAKE PARK

UPDATED: MAY 1989

TRAIL BEGINNING ELEVATION:
10,200 ft.

TRAIL ENDING ELEVATION:
10,700 ft.

ACCESS:
Access from the South Park District is from the Brookside-McCurdy Trail and the Wigwam Trail. The Lake Park Trail, much of which is on the South Platte District, extends from Hankins Pass 4 miles north and west to the Brookside-McCurdy Trail.

ATTRACTIONS:

Attractions: This is a high country route offering several outstanding views. Lake Park is a high elevation meadow surrounded by majestic rock formations and the snags of a ghost forest remain from the effects of a fire.

This is a great area for viewing alpine columbine a high-altitude miniature of the state flower. Back country water is available. Water is also, available at the Twin Eagles and Gooseneck Trailheads and Spruce Grove Campground.

USE: Moderate

RECOMMENDED SEASON:

SPRING	SUMMER	FALL	WINTER
	■	■	

NARRATIVE:

TRAIL PROFILE (ONE-WAY):

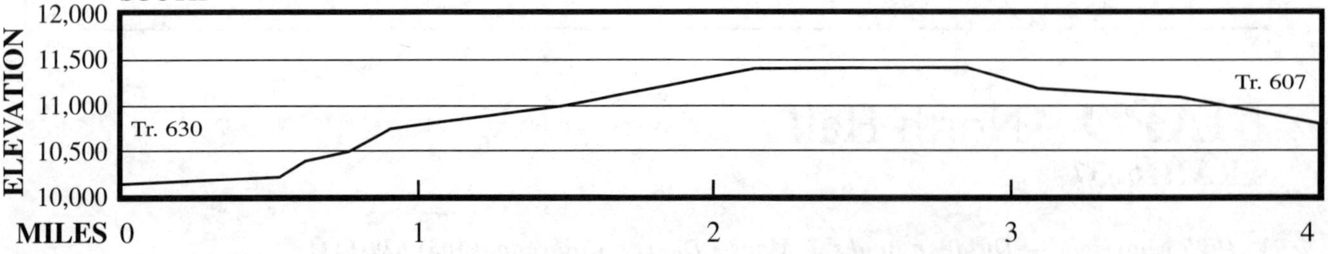

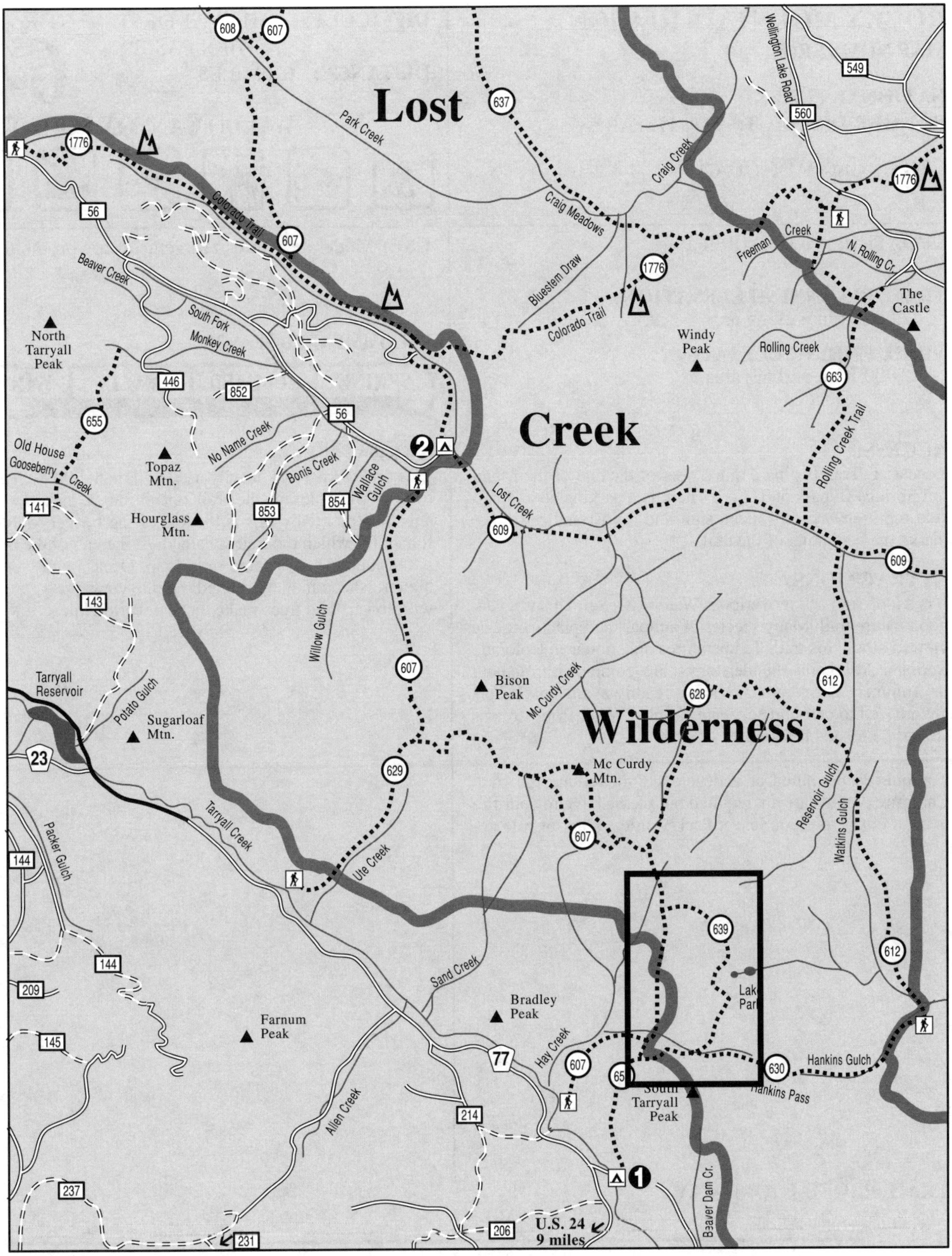

MAP 9

TRAIL #639

ROCKY MOUNTAIN REGION

MAP NUMBER(S): 16

NATIONAL FOREST: PIKE
RANGER DISTRICT: SOUTH PARK

USGS: CASCADE QUAD

DIFFICULTY: STEEP AT FIRST, THEN EASIER
DISTANCE: 6.5 MILES

TRAIL 640

WALDO CANYON LOOP TRAIL

UPDATED: NOVEMBER 1991

TRAIL BEGINNING ELEVATION:
7,000 ft. in parking area

TRAIL ENDING ELEVATION:
7,000 ft. in parking area

ACCESS:
Access: 1. Trail begins 2 miles west of the end of the Manitou Springs Bypass on U.S. 2. There is a parking area alongside the highway. A wooden sign and a post and rail fence marks the beginning of the trail.

ATTRACTIONS:
It is a loop trail and it overlooks Waldo Canyon. This is a very good nature trail. Many species of animals and plants may be viewed along this trail. Temperatures are similar to Colorado Springs. Afternoon thunderstorms are common throughout the summer. Drinking water is not readily available. Hikers are advised to carry an adequate supply. Water from streams should be treated before use.

Camping Is permitted at undeveloped sites along the trail. Campfire permits are not required but please be Sure your fire is dead out before you leave. Pack it in/Pack it Out rule applies.

USE: Moderate, Motorized vehicles are prohibited.

RECOMMENDED SEASON:

SPRING	SUMMER	FALL	WINTER

NARRATIVE:
This trail is the most heavily used trail on the Pikes Peak District . Two miles in the trail opens into a meadow near a stream. Reportedly this is the site of the former Waldo Hog Ranch for which the trail is named. At the end of the meadow is the junction of a three mile loop. Occasionally, deer and bighorn sheep may be spotted, but the view, birds, wildflowers and trees make the trip worthwhile.

TRAIL PROFILE (ONE-WAY):

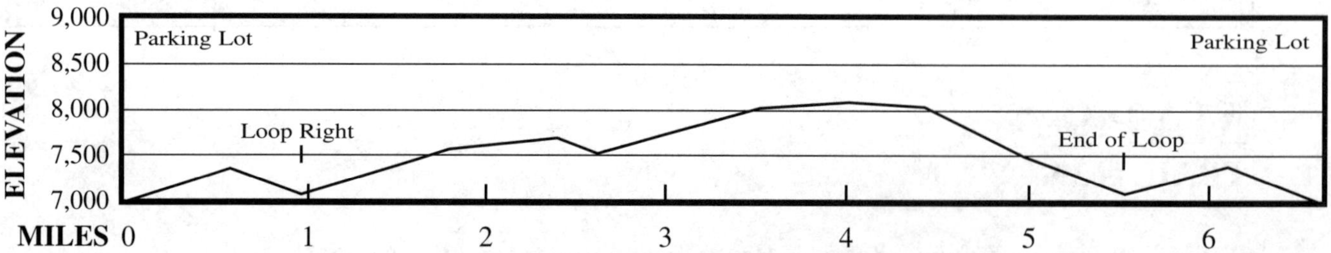

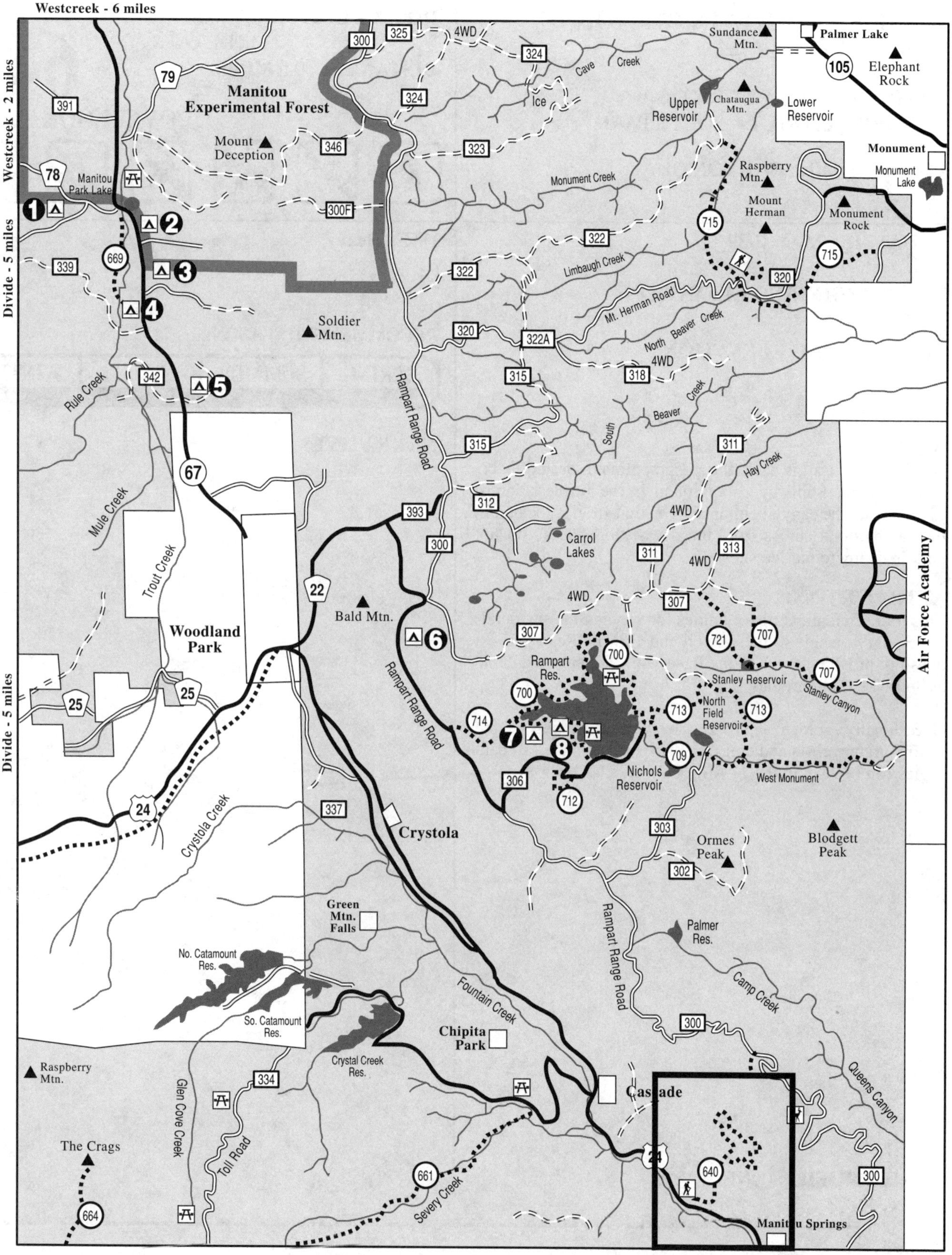

MAP 16
TRAIL #640

ROCKY MOUNTAIN REGION
MAP NUMBER(S): 14

NATIONAL FOREST: PIKE
RANGER DISTRICT: SOUTH PARK

USGS: 11-MILE CANYON QUAD

DIFFICULTY: EASY TO MODERATE
DISTANCE: 0.5 MILES

TRAIL 641
OVERLOOK TRAIL

UPDATED: MAY 1989

TRAIL BEGINNING ELEVATION:
8,500 ft.

TRAIL ENDING ELEVATION:
8,300 ft.

ACCESS:
Access to this trail is at Spillway Campground located in 11-Mile Canyon. Spillway campground is the furthest campground from the canyon entrance. The trail-head is located in the campground across from the pay station. Please do not park in campground fee sites.

ATTRACTIONS:
This trail offers great opportunities for viewing the surrounding country outside of the canyon and 11-Mile Reservoir. It does not provide access to the Reservoir. Water is available at Spillway Campground.

Occasionally, golden eagles can be viewed nesting in the nearby granite cliffs and crags, In the spring and late fall bald eagles can be seen in the 11-Mile Canyon area.

USE: Heavy

RECOMMENDED SEASON:

SPRING	SUMMER	FALL	WINTER

NARRATIVE:

TRAIL PROFILE (ONE-WAY):

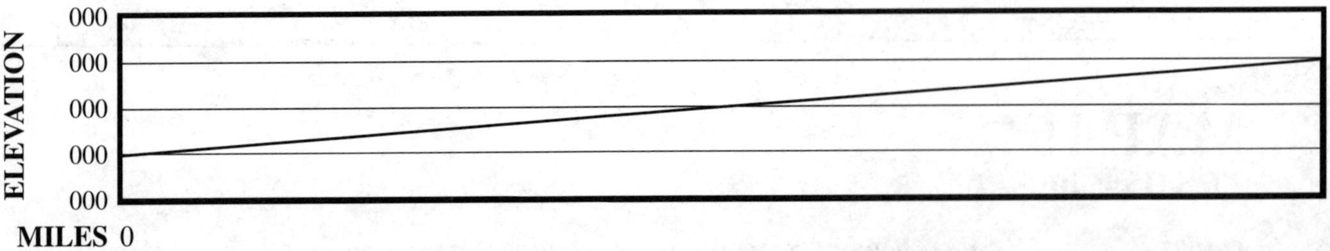

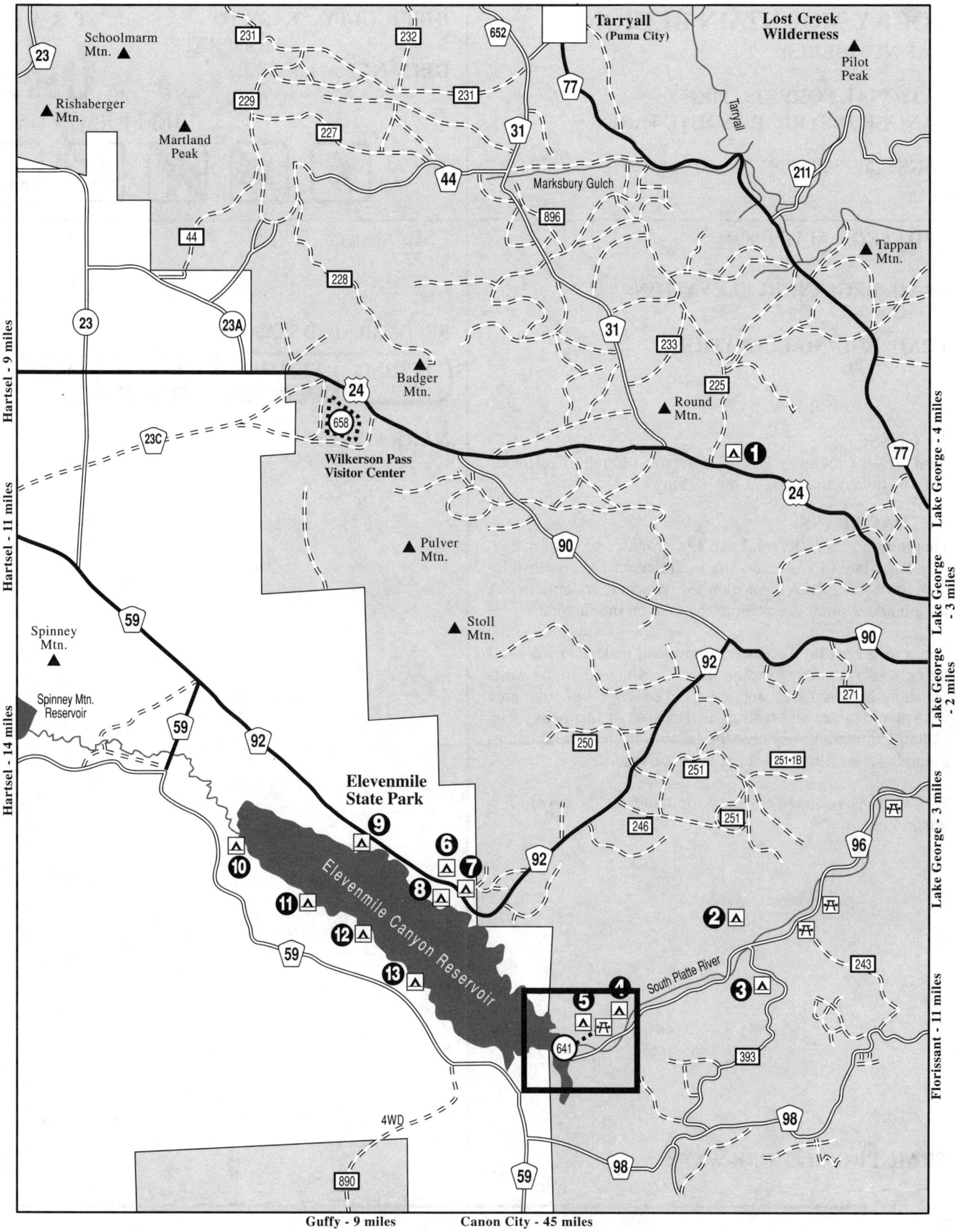

MAP 14

TRAIL #641

ROCKY MOUNTAIN REGION

MAP NUMBER(S): 2

NATIONAL FOREST: PIKE
RANGER DISTRICT: SOUTH PARK

USGS: JEFFERSON QUAD

DIFFICULTY: EASY TO MODERATE
DISTANCE: 1.5 MILES

TRAIL 642
JEFFERSON LAKE

UPDATED: MAY 1989

TRAIL BEGINNING ELEVATION:
10,600 ft.

TRAIL ENDING ELEVATION:
10,500 ft.

ACCESS:
Turn west of highway 285 in Jefferson. Drive to Jefferson Lake (8 miles) and park, at the parking lot.

ATTRACTIONS:
This trail circles Jefferson Lake and provides access for fishing. The State of Colorado stocks Jefferson Lake with rainbow trout. The lake is surrounded by scenic mountains. A stream and a small brook enter the lake on the far side.

The trail is best on the eastern shore and makes a nice stroll for persons who do not ordinarily hike. The trail on the western shore is more rocky. Jefferson Lake is a very popular area with picnic tables and restrooms. Because of this heavy use, camping is not allowed near the lake itself but several campgrounds are available on the way to the lake.

Well water is available at the campgrounds on the way to Jefferson Lake.

USE: Moderate

RECOMMENDED SEASON:

SPRING	SUMMER	FALL	WINTER

NARRATIVE:

TRAIL PROFILE (ONE-WAY):

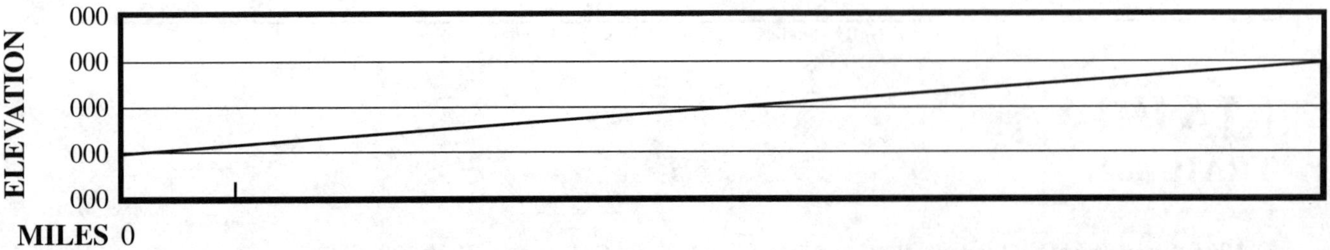

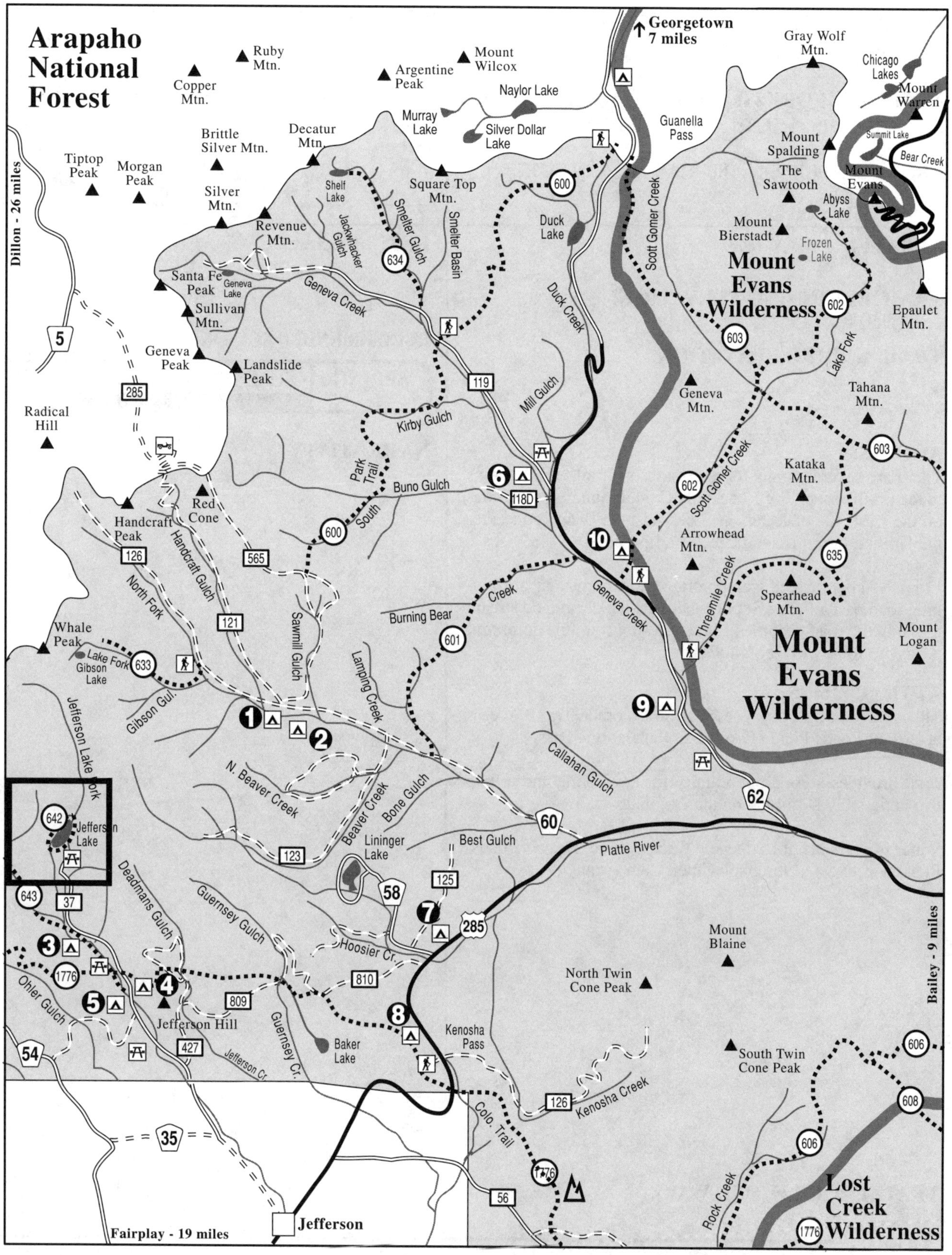

MAP 2

TRAIL #642

ROCKY MOUNTAIN REGION

MAP NUMBER(S): 1 and 2

NATIONAL FOREST: PIKE
RANGER DISTRICT: SOUTH PARK

USGS: JEFFERSON, BOREAS PASS QUADS

DIFFICULTY: MODERATE

DISTANCE: 5.0 MILES

TRAIL 643
WEST JEFFERSON

UPDATED: MAY 1989

TRAIL BEGINNING ELEVATION:
9,950 ft.

TRAIL ENDING ELEVATION:
11,585 ft.

ACCESS:
There are several points for accessing this trail. Trail begins in the northwest end of Lodgepole Campground. It continues on the west side of Jefferson Creek past the Beaverpond Picnic Ground into Jefferson Creek Campground.

From Jefferson Creek Campground trail follows an old logging road past a Forest Service gate. At the developed stream crossing the road becomes a hiking trail. From this point trail is easy to follow to Georgia Pass.

ATTRACTIONS:
This is a good trail for accessing the Colorado Trail. Access at Beaver Ponds Picnic Ground and at Georgia Pass.

Trail provides a good opportunity for day hiking and a loop trip is possibly by hiking on the Colorado Trail.

Water is available at Lodgepole Campground. Beaver Ponds Picnic Ground and Jefferson Creek Campground.

USE: Moderate

RECOMMENDED SEASON:

SPRING	SUMMER	FALL	WINTER

NARRATIVE:

TRAIL PROFILE (ONE-WAY):

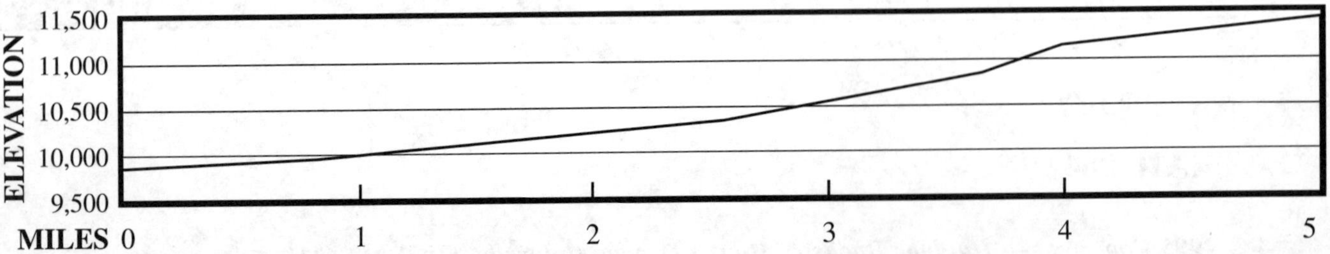

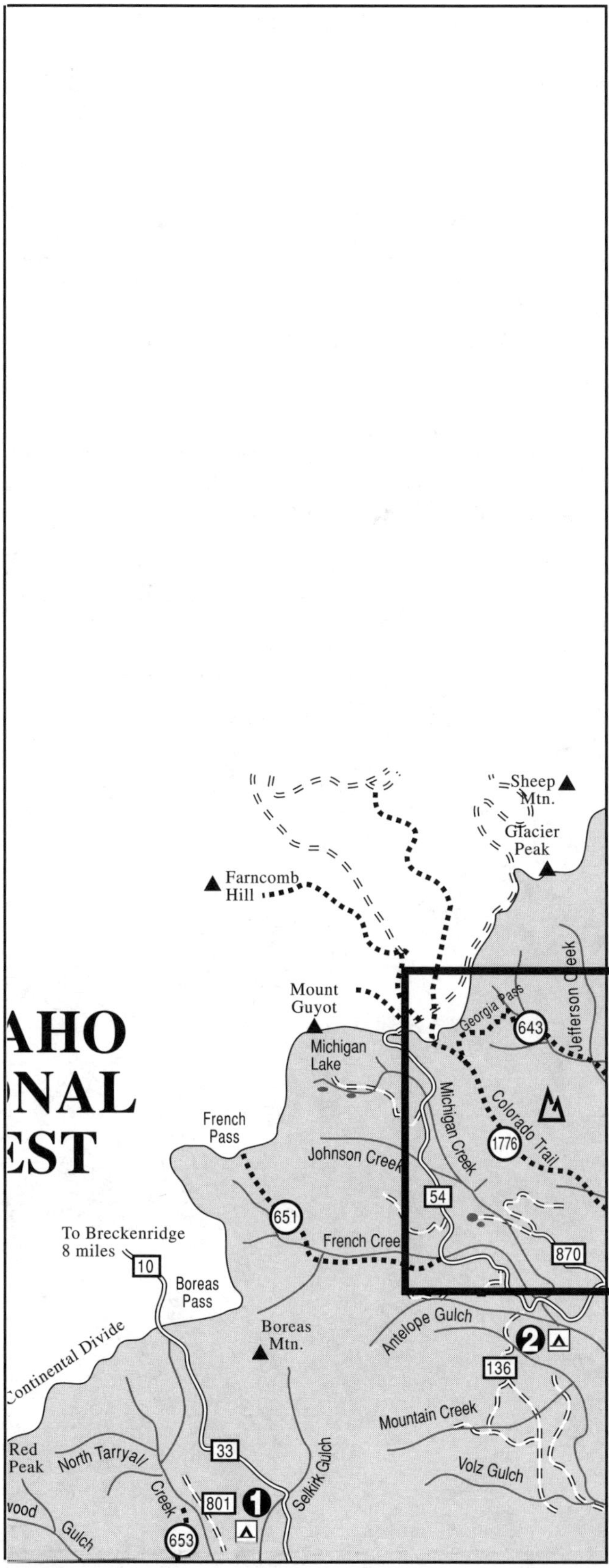

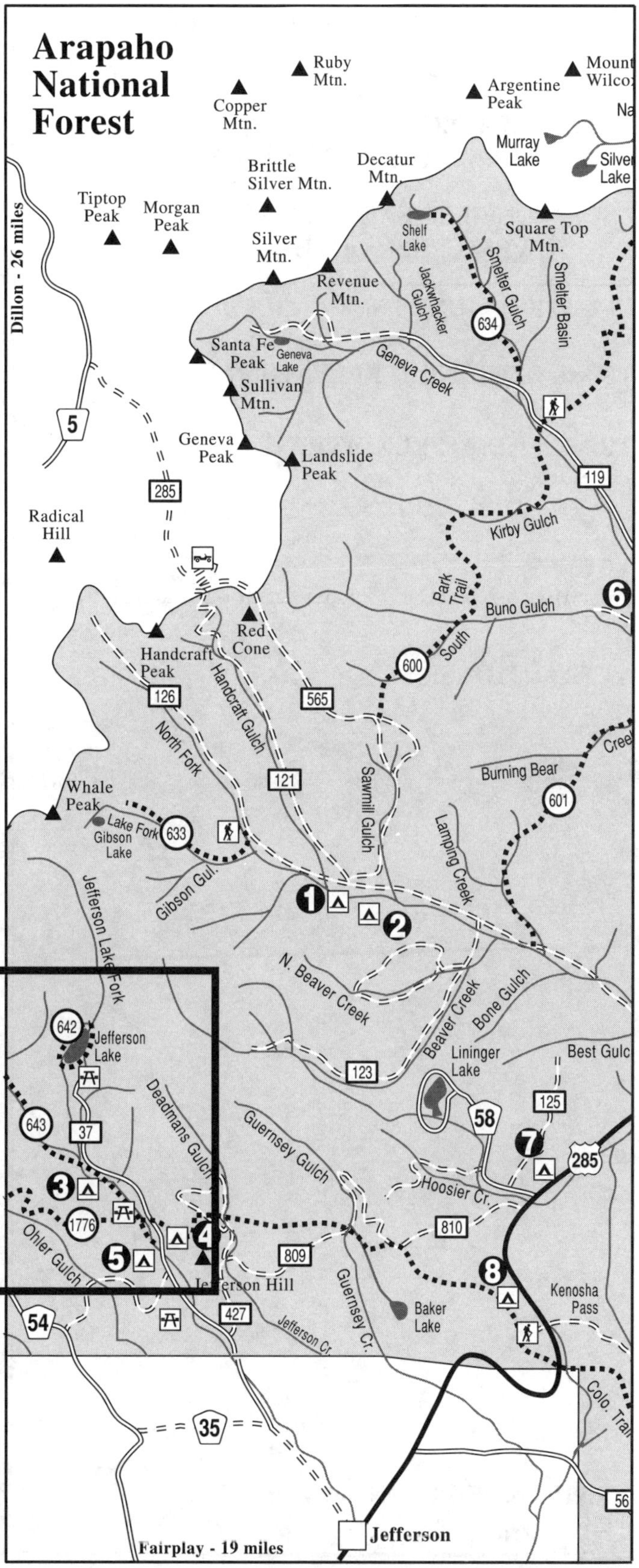

MAP 1 - East Half

TRAIL #643

MAP 2 - West Half

ROCKY MOUNTAIN REGION

MAP NUMBER(S): 10 and 11

NATIONAL FOREST: PIKE
RANGER DISTRICT: SOUTH PLATTE

USGS: DEVILS HEAD, DAKAN MTN., WESTCREEK QUADS

DIFFICULTY: INTERMEDIATE

DISTANCE: 6.2 MILES

TRAIL 649
RAMPART RANGE

UPDATED: JUNE 1988

TRAIL BEGINNING ELEVATION:
8,550 ft.

TRAIL ENDING ELEVATION:
8,505 ft.

ACCESS:
From Hwy 67 proceed 15 miles south on Rampart Range Road.

ATTRACTIONS:

USE: Moderate
HIKING, BIKING, MOTORCYCLING, ATV's

RECOMMENDED SEASON:

SPRING | SUMMER | FALL | WINTER

NARRATIVE:
A beautiful loop trail that accesses Trail 650 and Trout Creek. Trout Creek has excellent fishing opportunities. During April and May, Trout Creek may be too deep to cross causing the trail to become two dead end trails instead of one loop.

TRAIL PROFILE (ONE-WAY):

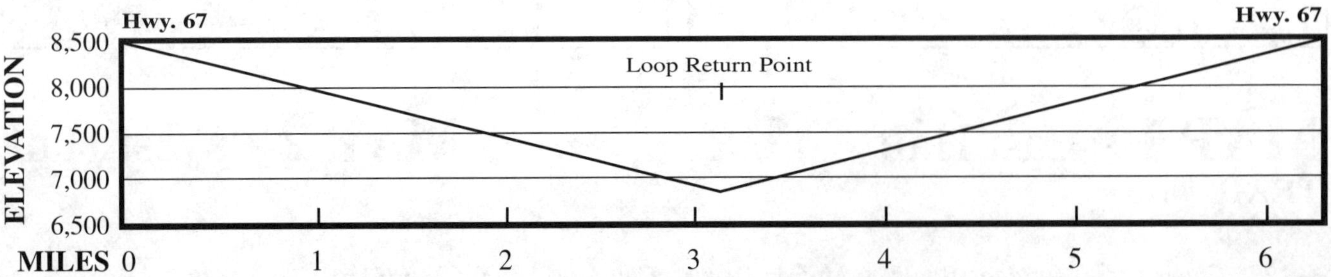

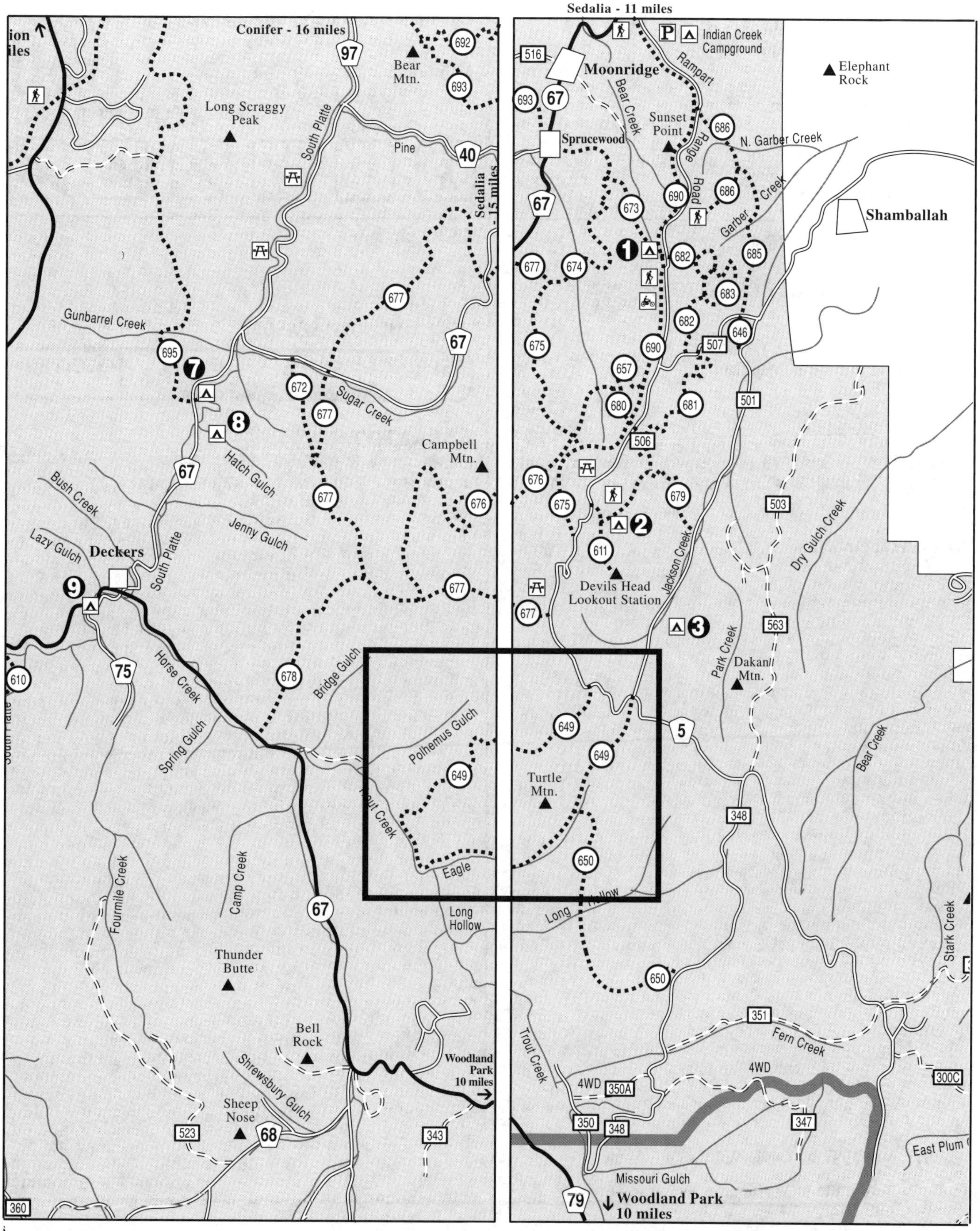

MAP 10 - East Half

TRAIL #649

MAP 11 - West Half

ROCKY MOUNTAIN REGION
MAP NUMBER(S): 11

NATIONAL FOREST: PIKE
RANGER DISTRICT: SOUTH PLATTE

USGS: DEVILS HEAD, DAKAN MTN., WESTCREEK QUADS

DIFFICULTY: INTERMEDIATE

DISTANCE: 5.5 MILES

TRAIL 650

RAMPART RANGE

UPDATED: JUNE 1988

TRAIL BEGINNING ELEVATION:
8,600 ft. Forest Road 348

TRAIL ENDING ELEVATION:
8,000 ft. junction with trail 649

ACCESS:
From Hwy 67 proceed 18 miles south on Rampart Range Road to Forest Road 348 then 4 miles south on 348 to trailhead.

ATTRACTIONS:

USE: Moderate

RECOMMENDED SEASON:

SPRING	SUMMER	FALL	WINTER

NARRATIVE:
An intermediate trail that in combination with trail 649 becomes 1 or 2 loop trails.

TRAIL PROFILE (ONE-WAY):

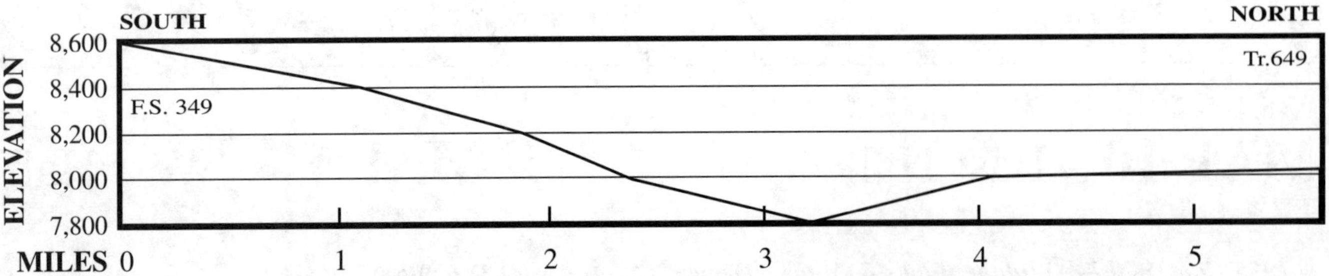

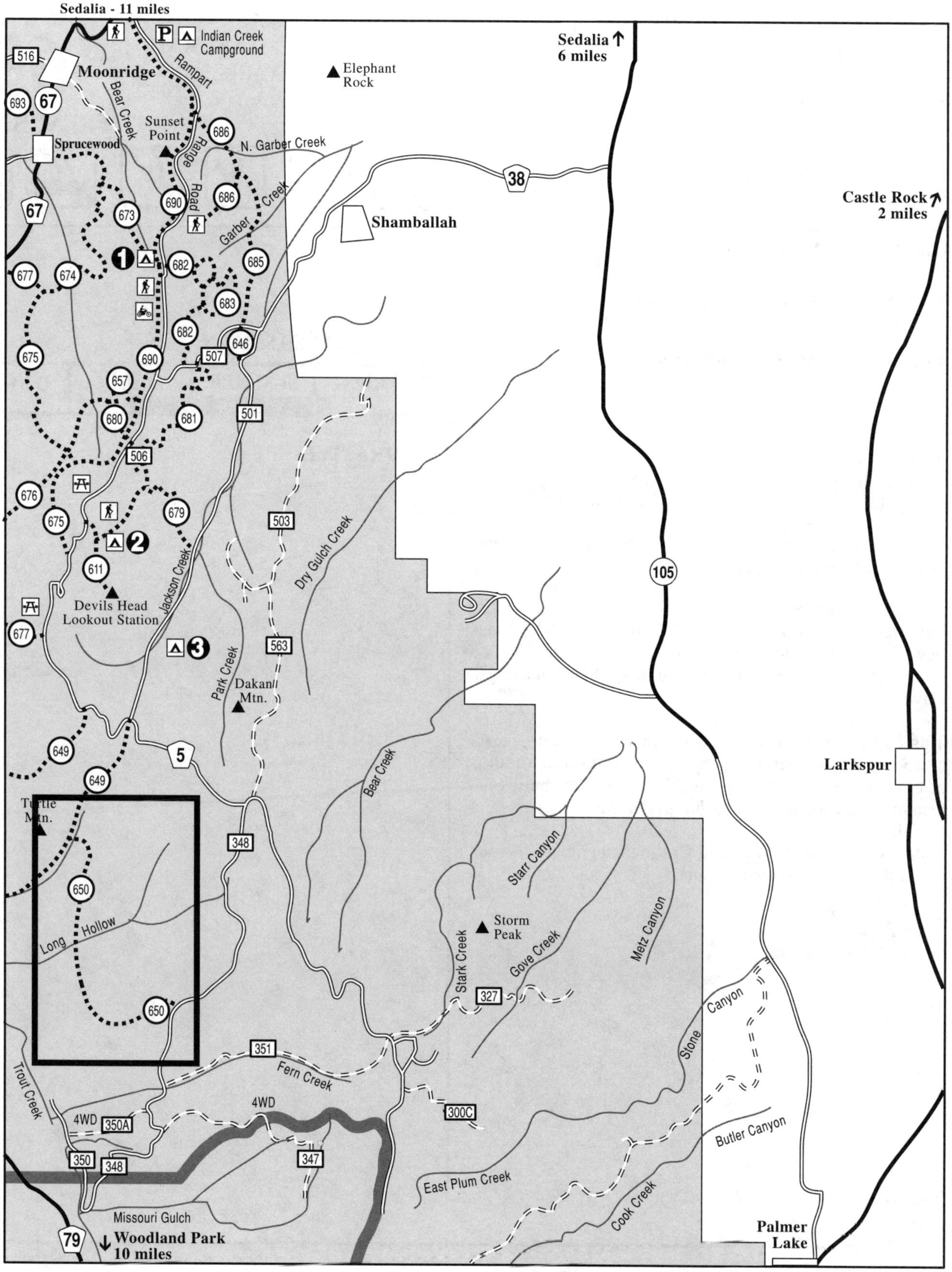

MAP 11
TRAIL #650

ROCKY MOUNTAIN REGION
MAP NUMBER(S): 1

NATIONAL FOREST: PIKE
RANGER DISTRICT: SOUTH PARK

USGS: BOREAS PASS QUAD

DIFFICULTY: EASY TO MODERATE
DISTANCE: 4.5 MILES

TRAIL 651
FRENCH PASS

UPDATED: MAY 1989

TRAIL BEGINNING ELEVATION:
11,000 ft.

TRAIL ENDING ELEVATION:
12,000 ft.

ACCESS:
The French Pass Trail begins off of the Michigan Creek Road (FDR 400). Continue on this road toward Georgia Pass. Drive 2 1/4 miles pass Michigan Campground to a point where the road crosses French Creek. Park in a flat area just before reaching this crossing.

ATTRACTIONS:
This trail provides access to the Continental Divide, Black Powder Pass Trail and Boreas Pass Road. French Pass is a Continental Divide crossing between Bald Mountain and Mt. Guyot.

This trail is not always easy to follow and not well signed. The view to the south as the hiker returns down the trail is across South Park to the Tarryall Mountains. This trail is not recommended for winter use due to the high avalanche potential.
Water is available in Michigan Creek campground. French Creek provides backcountry water.

USE: Moderate

RECOMMENDED SEASON:

SPRING	SUMMER	FALL	WINTER

NARRATIVE:

TRAIL PROFILE (ONE-WAY):

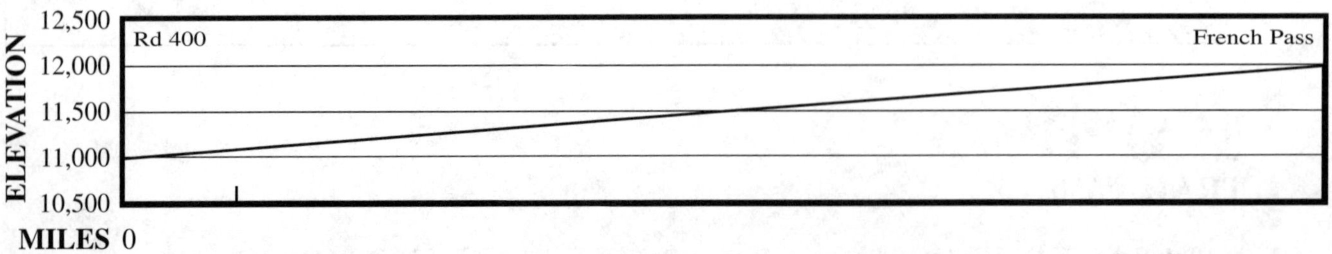

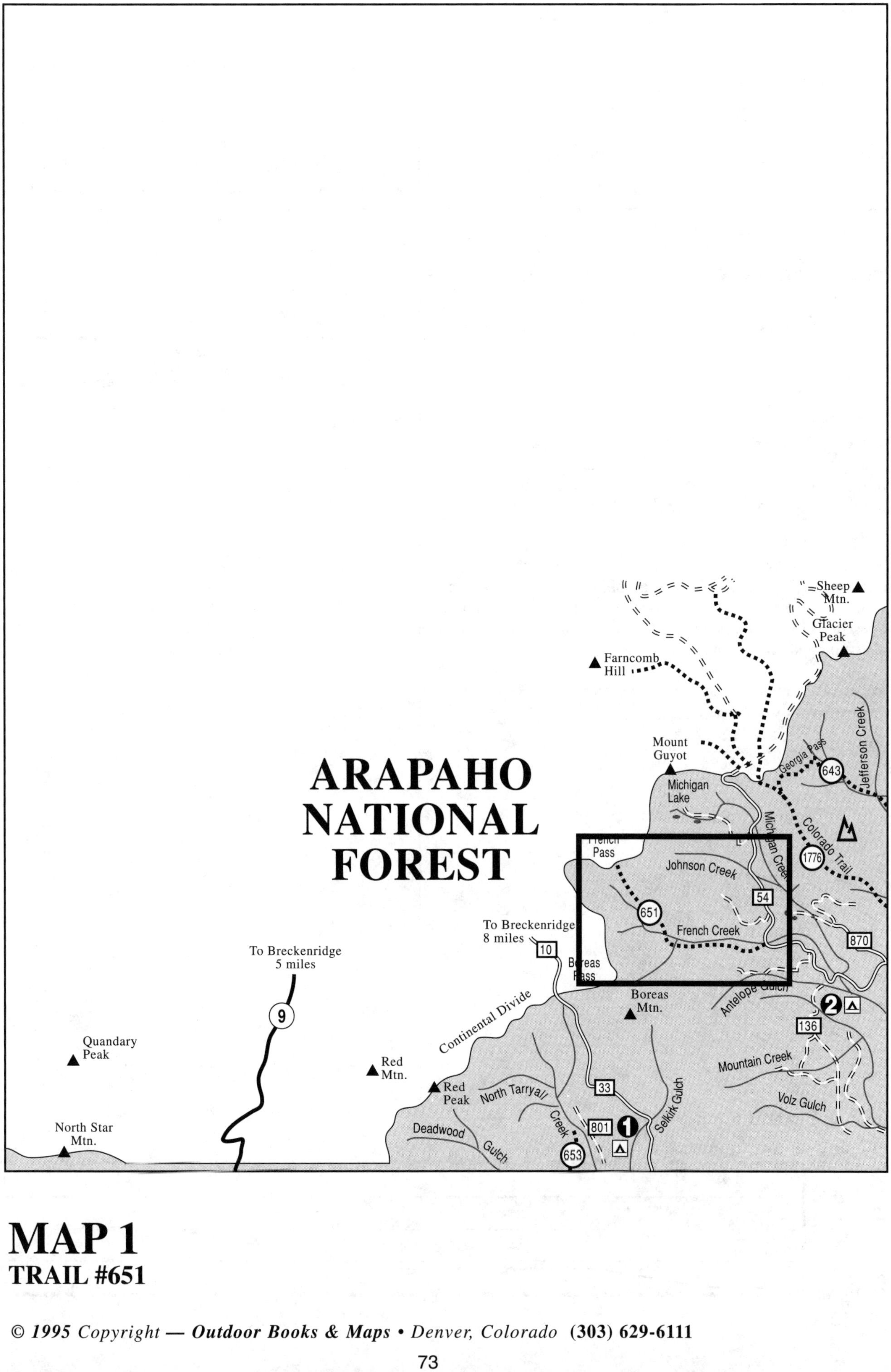

MAP 1
TRAIL #651

ROCKY MOUNTAIN REGION

MAP NUMBER(S): 1 and 7

NATIONAL FOREST: PIKE
RANGER DISTRICT: SOUTH PARK

USGS: COMO, BOREAS PASS QUADS

DIFFICULTY: MODERATE

DISTANCE: 8.0 MILES

TRAIL 653
GOLD DUST

UPDATED: MAY 1989

TRAIL BEGINNING ELEVATION:
10,000 ft.

TRAIL ENDING ELEVATION:
11,200 ft.

ACCESS:
The Gold Dust Trail begins off the Tarryall Road (FDR 405) in the town of Como. The trailhead is approximately 5.5 miles from US Highway 285. There is a wide spot for parking in the road approximately .03 miles past the trailhead.

ATTRACTIONS:
This trail provides access to Boreas Pass. Past the first 1/2 mile, the trail generally follows an old flume which diverted water from North Tarryall Creek and its tributaries to mining operations located in the upper Tarryall Drainage. This trail is also a popular cross-country ski trail.

Water is not available at the trail heads or at Selkirk Campground.

USE: Moderate

RECOMMENDED SEASON:

SPRING	SUMMER	FALL	WINTER

NARRATIVE:

TRAIL PROFILE (ONE-WAY):

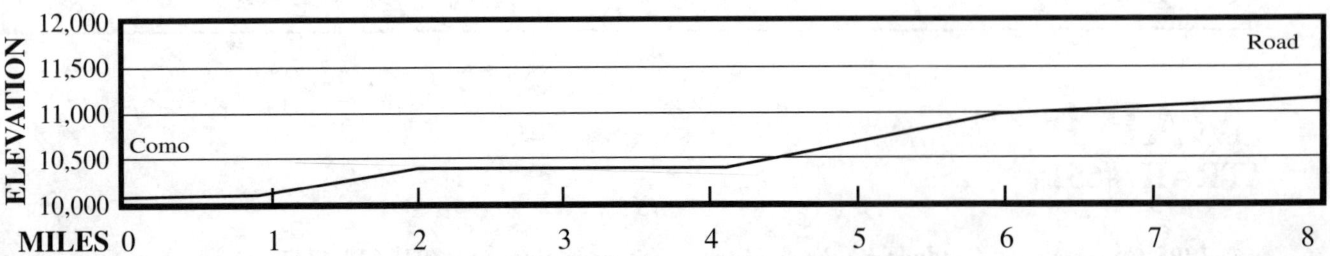

MAP 1 - South Third

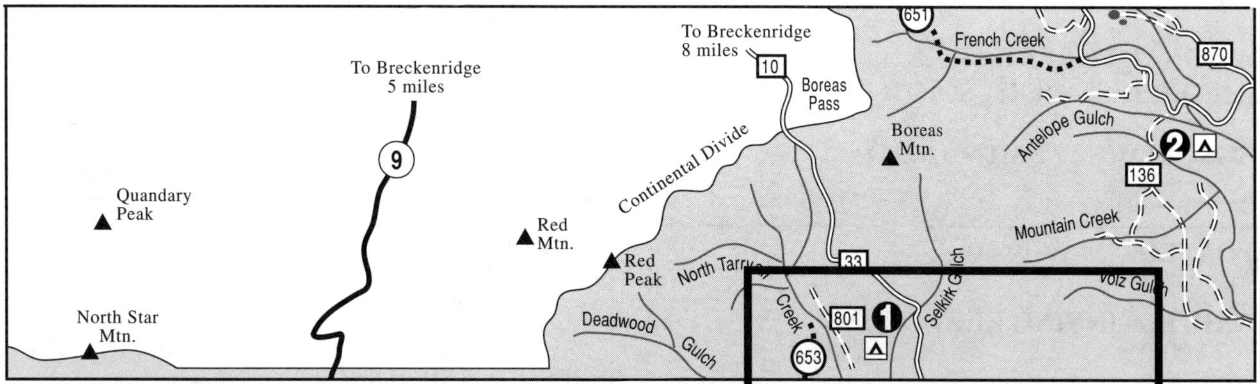

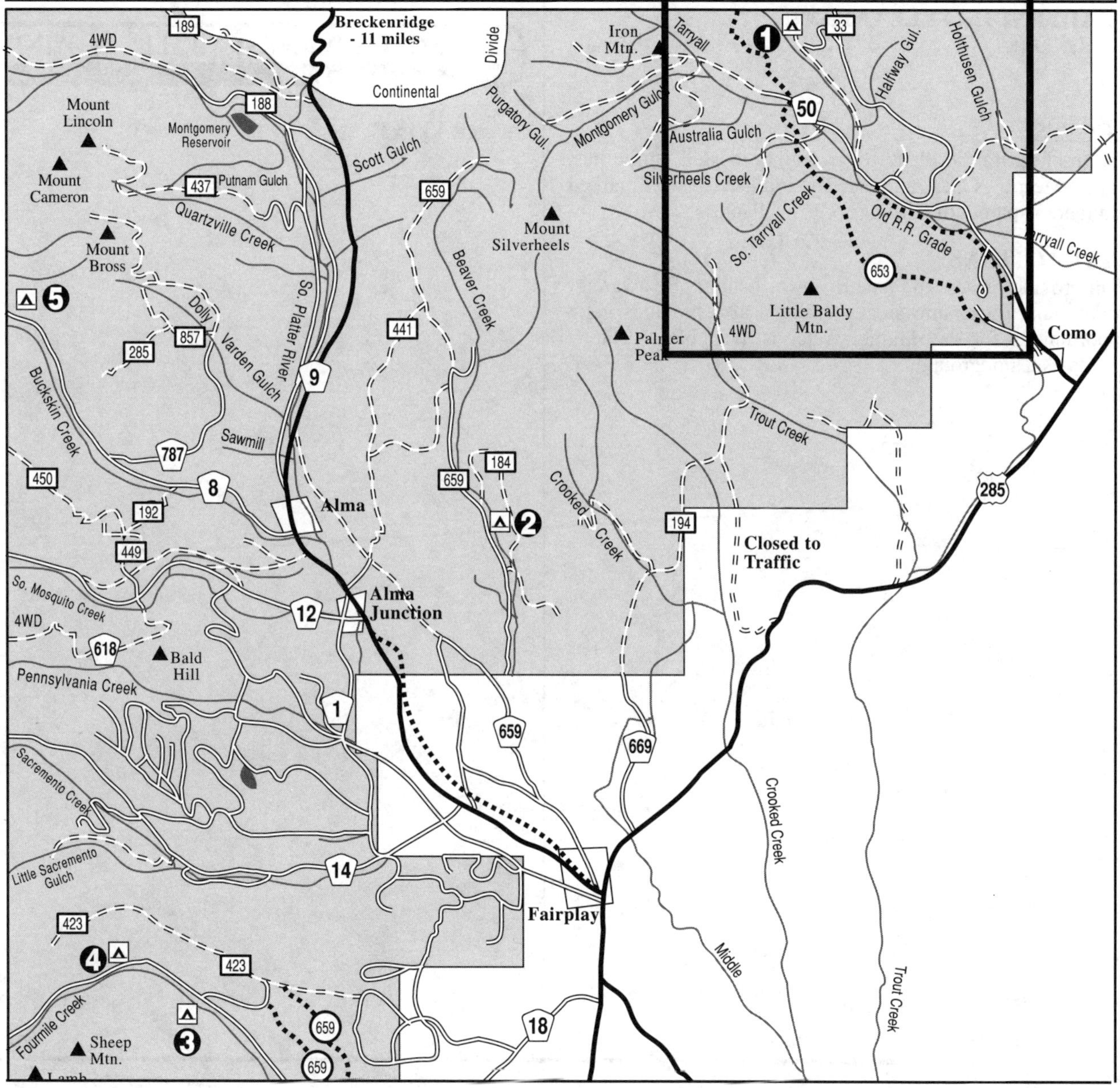

MAP 7 - North Two-Thirds

TRAIL #653

ROCKY MOUNTAIN REGION

MAP NUMBER(S): 15

NATIONAL FOREST: PIKE
RANGER DISTRICT: SOUTH PARK

USGS: HACKETT MTN. QUAD

DIFFICULTY: EASY TO MODERATE
DISTANCE: 3.5 MILES

TRAIL 654 PLATTE RIVER

UPDATED: MAY 1989

TRAIL BEGINNING ELEVATION:
7,800 ft.

TRAIL ENDING ELEVATION:
7,500 ft.

ACCESS:
The trailhead to the Platte River Trail is located 1 mile past Happy Meadows Campground. The trailhead is well marked and there is plenty of parking space available.

ATTRACTIONS:
Trail provides access to Platte River for fishing. Be aware of private land ownership areas. This trail also provides good opportunities for day hiking. Water is available at Happy Meadows Campground.

USE: Moderate

RECOMMENDED SEASON:

SPRING	SUMMER	FALL	WINTER

NARRATIVE:

TRAIL PROFILE (ONE-WAY):

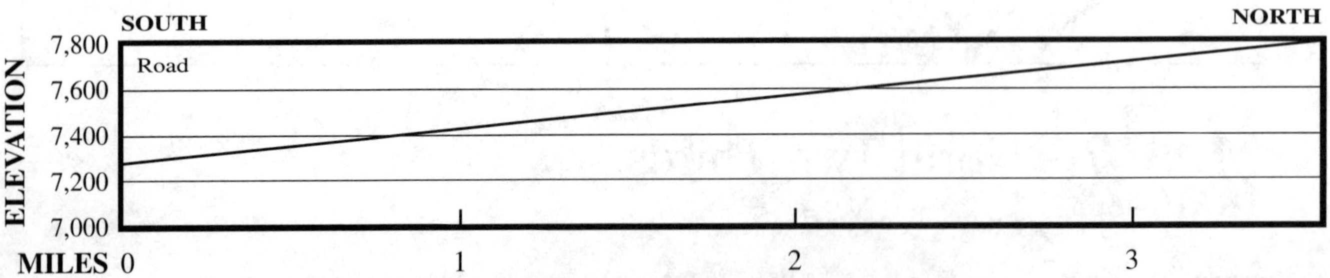

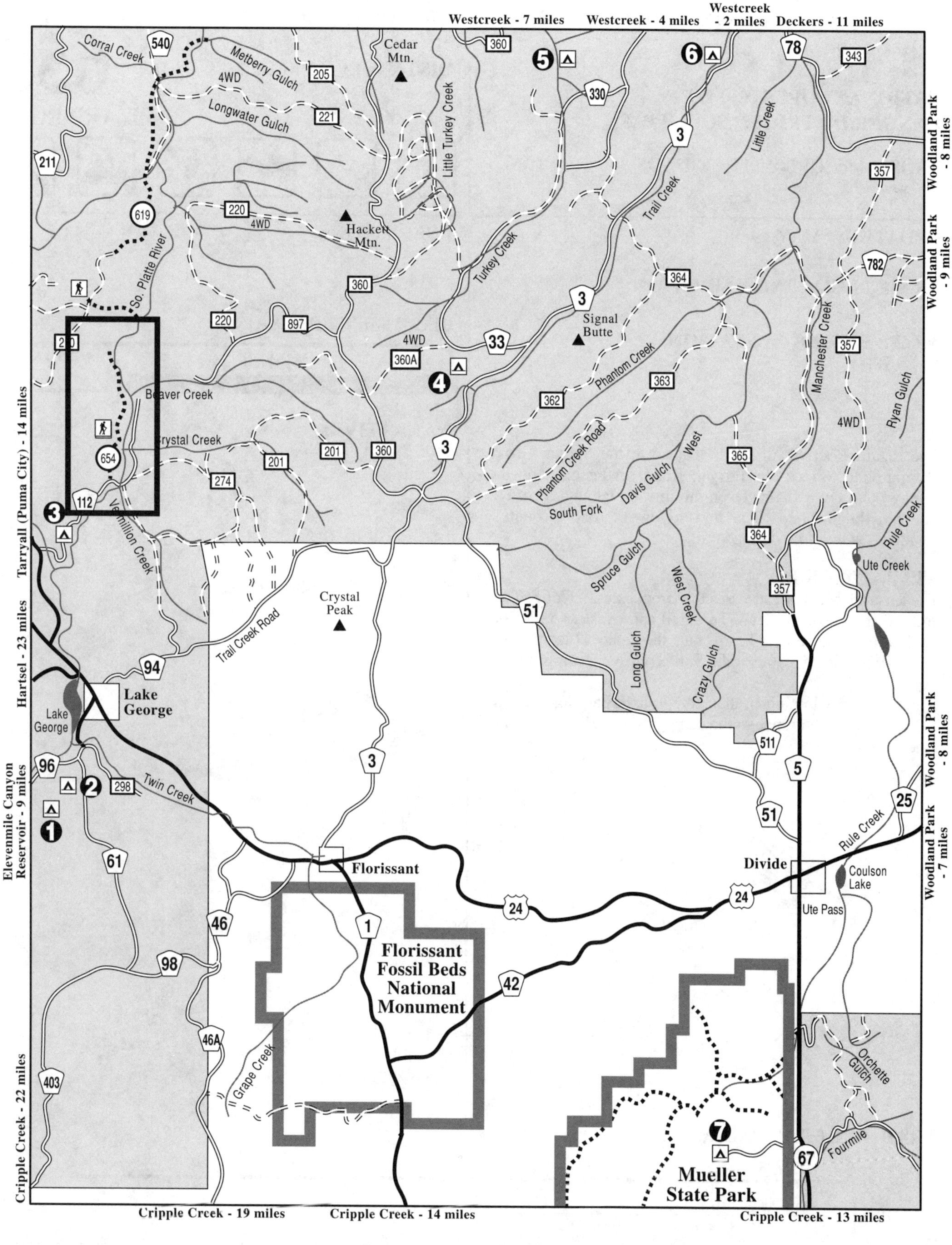

MAP 15

TRAIL #654

ROCKY MOUNTAIN REGION

MAP NUMBER(S): 9

NATIONAL FOREST: PIKE
RANGER DISTRICT: SOUTH PARK

USGS: McCURDY MTN. QUAD

DIFFICULTY: EASY TO MODERATE
DISTANCE: 2.3 MILES

TRAIL 655
LIZARD ROCK

UPDATED: MAY 1989

TRAIL BEGINNING ELEVATION:
8,500 ft.

TRAIL ENDING ELEVATION:
8,700 ft.

USE: Moderate

RECOMMENDED SEASON:

SPRING	SUMMER	FALL	WINTER

ACCESS:
The Lizard Rock Trailhead is located at the Spruce Grove Campground, off of the Tarryall Road. After entering the campground cross the river on the foot bridge and turn left. The trail begins here. The trail follows a road through a meadow leading to a defined trail.

ATTRACTIONS:
The Lizard Rock Trail accesses the Brookside-McCurdy Trail and the Hankins Pass Trail. This trail approaches a rock formation that resembles a lizard, thus the name. "Lizard Rock Trail". Water is available at the Spruce Grove Campground.

If you are not camping in the campground, parking is available outside the entrance to the campground near the wooden fence.

NARRATIVE:

TRAIL PROFILE (ONE-WAY):

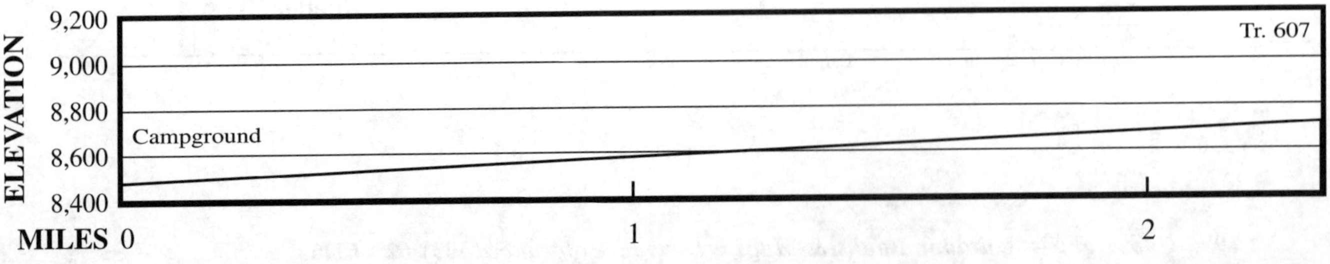

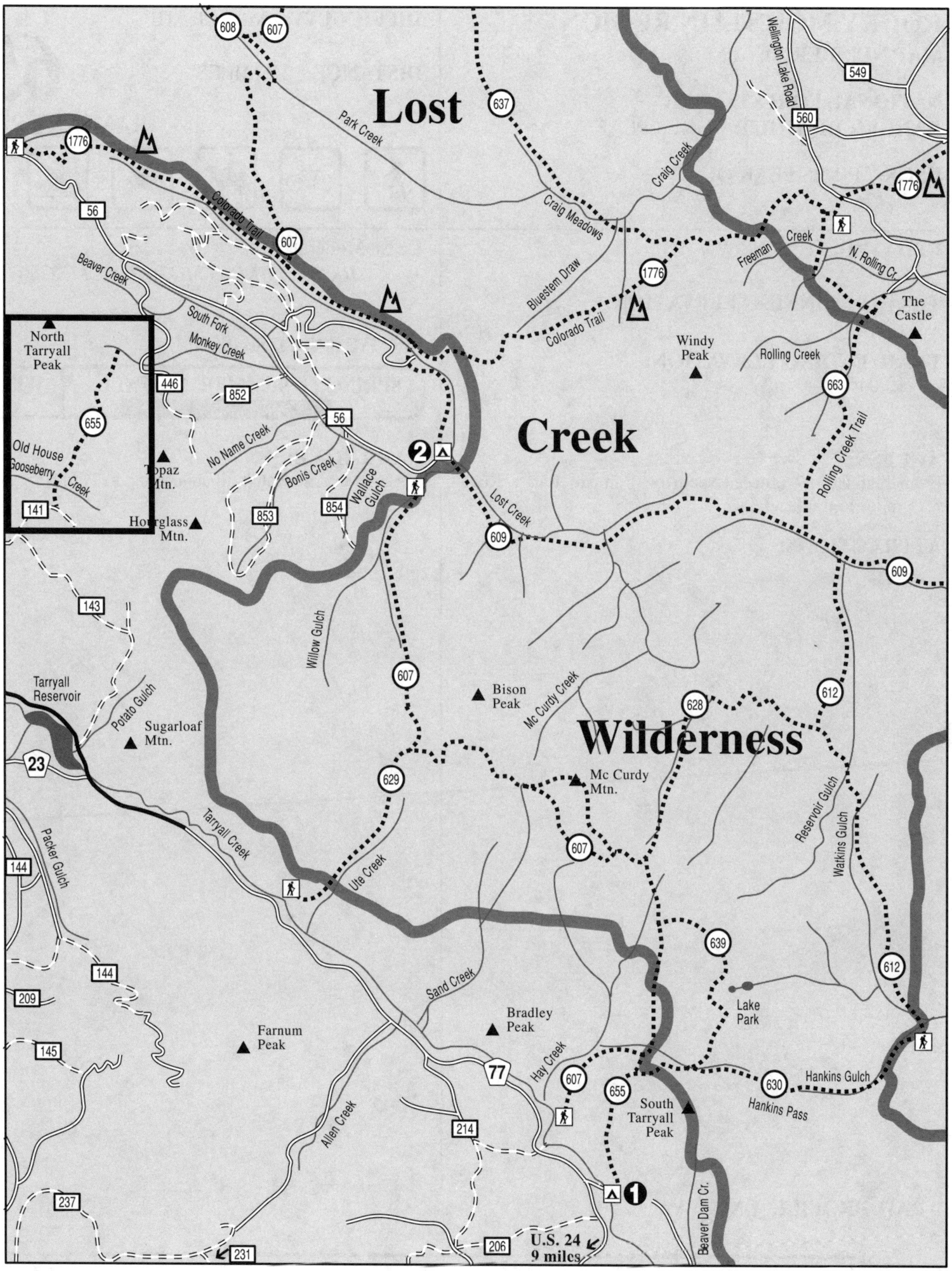

MAP 9

TRAIL #655

ROCKY MOUNTAIN REGION
MAP NUMBER(S): 11

NATIONAL FOREST: PIKE
RANGER DISTRICT: PIKES PEAK

USGS: PIKES PEAK QUAD

DIFFICULTY: MODERATE

DISTANCE: 2.7 MILES

TRAIL 657
RAMPART RANGE

UPDATED: UNKNOWN

TRAIL BEGINNING ELEVATION:
10,500 ft.

TRAIL ENDING ELEVATION:
12,046 ft.

ACCESS:
From highway 67 proceed south on Rampart Range Road 5.75 miles to trailhead 657.

ATTRACTIONS:

USE: Moderate
HIKING, BIKING MOTORCYCLING, ATV's

RECOMMENDED SEASON:

SPRING	SUMMER	FALL	WINTER

NARRATIVE:
A scenic intermediate trail connecting to Trail 675.

TRAIL PROFILE (ONE-WAY):

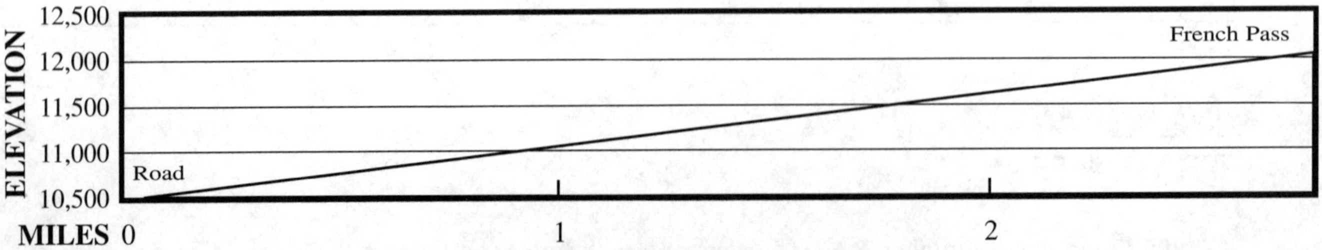

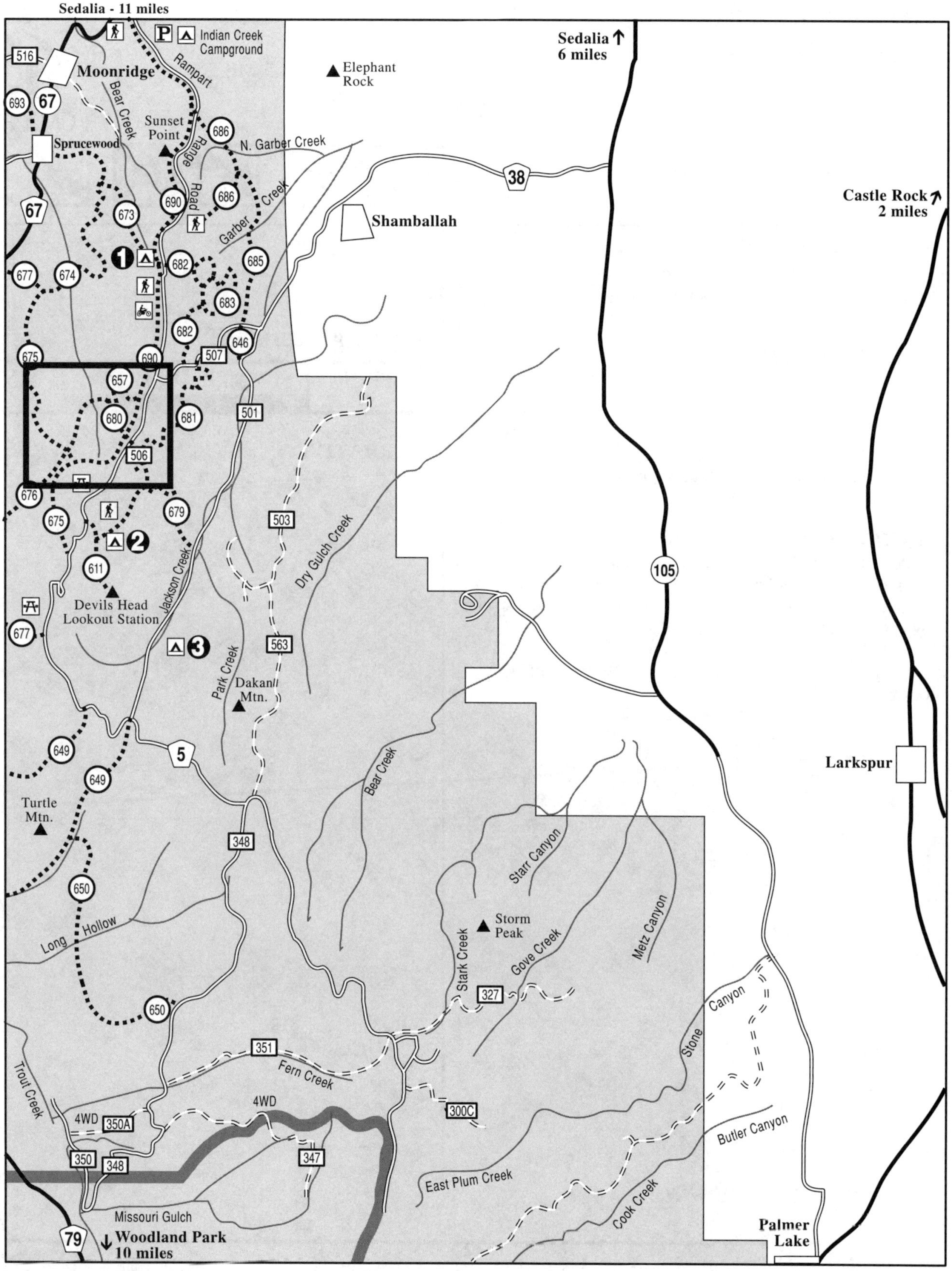

MAP 11

TRAIL #657

ROCKY MOUNTAIN REGION

MAP NUMBER(S): 14

NATIONAL FOREST: PIKE
RANGER DISTRICT: SOUTH PARK

USGS: GLENTIVAR QUAD

DIFFICULTY: EASY

DISTANCE: 1.5 MILES

TRAIL 658 PUMA POINT

UPDATED: MAY 1989

TRAIL BEGINNING ELEVATION:
9,500 ft.

TRAIL ENDING ELEVATION:
9,500 ft.

ACCESS:

The Puma Point trail begins at the Wilkerson Pass Visitor center off of U.S. Highway 24.

ATTRACTIONS:

While visiting Wilkerson Pass folks should take advantage of this interpretive trail. Outstanding scenic views of South Park the Collegiate Peaks and the Mosquito Range would excite any camera buff.

Restrooms and water are available at the Wilkerson Pass Visitor Center.

USE: Moderate

RECOMMENDED SEASON:

SPRING	SUMMER	FALL	WINTER

NARRATIVE:

TRAIL PROFILE (ONE-WAY):

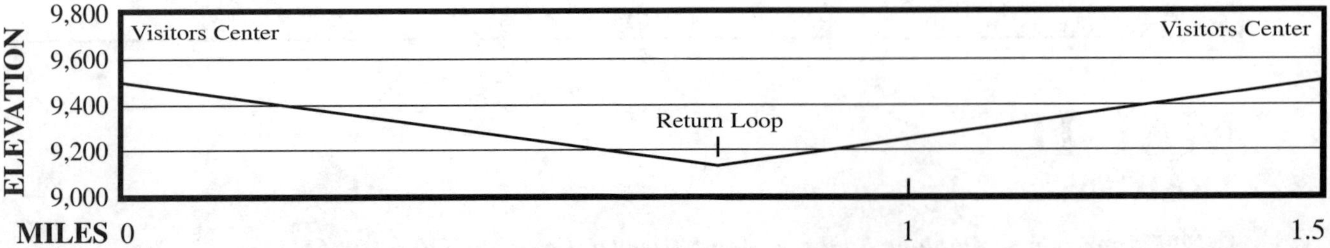

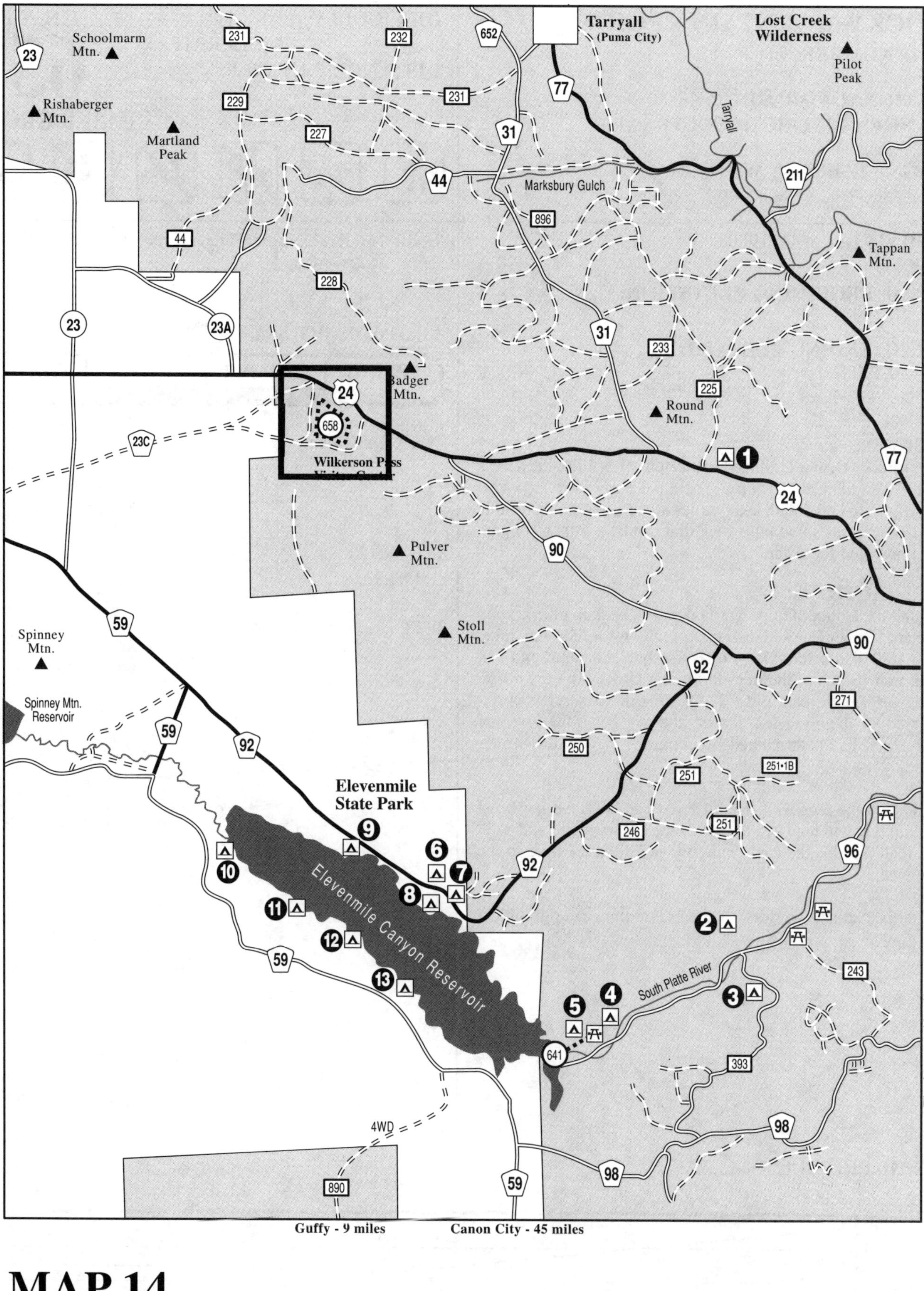

MAP 14

TRAIL #658

ROCKY MOUNTAIN REGION

MAP NUMBER(S): 7

NATIONAL FOREST: PIKE
RANGER DISTRICT: SOUTH PARK

USGS: FAIRPLAY WEST QUAD

DIFFICULTY: EASY TO MODERATE
DISTANCE: 1.5 MILES

TRAIL 659
LIMBER GROVE

UPDATED: MAY 1989

TRAIL BEGINNING ELEVATION:
10,750 ft.

TRAIL ENDING ELEVATION:
10,500 ft.

ACCESS:
The Limber Grove Trailhead is located off of FDR 2I, across from Fourmile Campground. There is a parking area provided. High-top boots with heavy soles are recommended for this trail. It crosses some talus rock that can be a little too much for thin-soled footwear.

ATTRACTIONS:
Hiking the Limber Grove Trail takes you back in time among ancient Limber Pines. Good opportunity for a short day hike. This trail also offers views of Horseshoe Mountain and Mt. Sherman. Some of the trees in Limber Grove are very possibly over 1,000 years old. Their struggle to survive is expressed in their twisted shapes and the beautiful grain of wood that has been etched by centuries of howling winter winds.

To extend the length of this hike you can continue northeast on trail that will lead you to Horseshoe Campground and walk the road back to the car. This loop adds about I mile to the distance.

Water is available at Horseshoe and Fourmile campgrounds.

USE: Moderate, horses are not recommended for this trail

RECOMMENDED SEASON:

SPRING	SUMMER	FALL	WINTER

NARRATIVE:

TRAIL PROFILE (ONE-WAY):

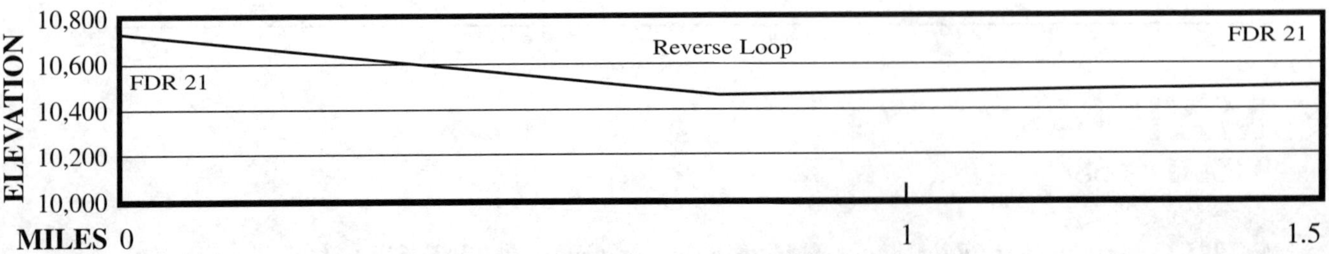

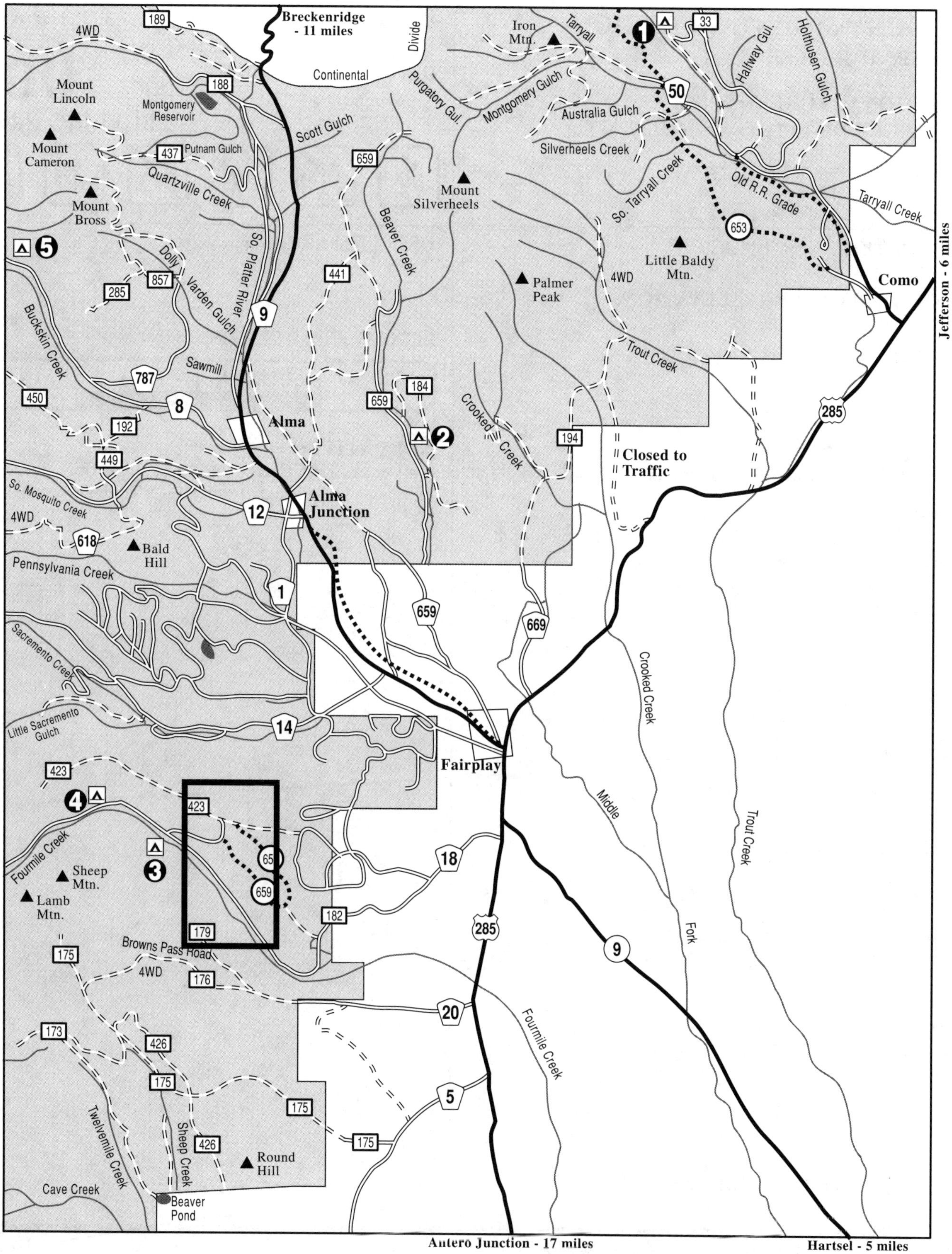

MAP 7
TRAIL #659

ROCKY MOUNTAIN REGION
MAP NUMBER(S): 9

NATIONAL FOREST: PIKE
RANGER DISTRICT: SOUTH PLATTE

USGS: WINDY PEAK QUAD

DIFFICULTY: EASY TO MODERATE
DISTANCE: 4.0 MILES

TRAIL 663
ROLLING CREEK

UPDATED: SEPTEMBER 1990

TRAIL BEGINNING ELEVATION:
8,200 ft. at Trailhead

TRAIL ENDING ELEVATION:
9,600 ft. at Trail 609

ACCESS:
Forest Service Road 543 to Rolling Creek turnoff.

ATTRACTIONS:
View, rock formations, trailhead parking, trail ends with no trace

USE: Light to Moderate

RECOMMENDED SEASON:

SPRING	SUMMER	FALL	WINTER

NARRATIVE:
Trail enters Lost Creek Wilderness approximately one-half mile south of Trailhead. Terminates at Trail 609 near Wigwam Park.

TRAIL PROFILE (ONE-WAY):

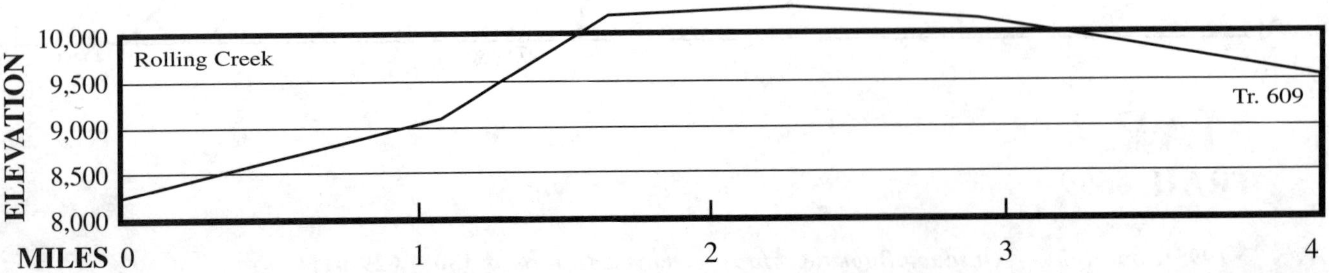

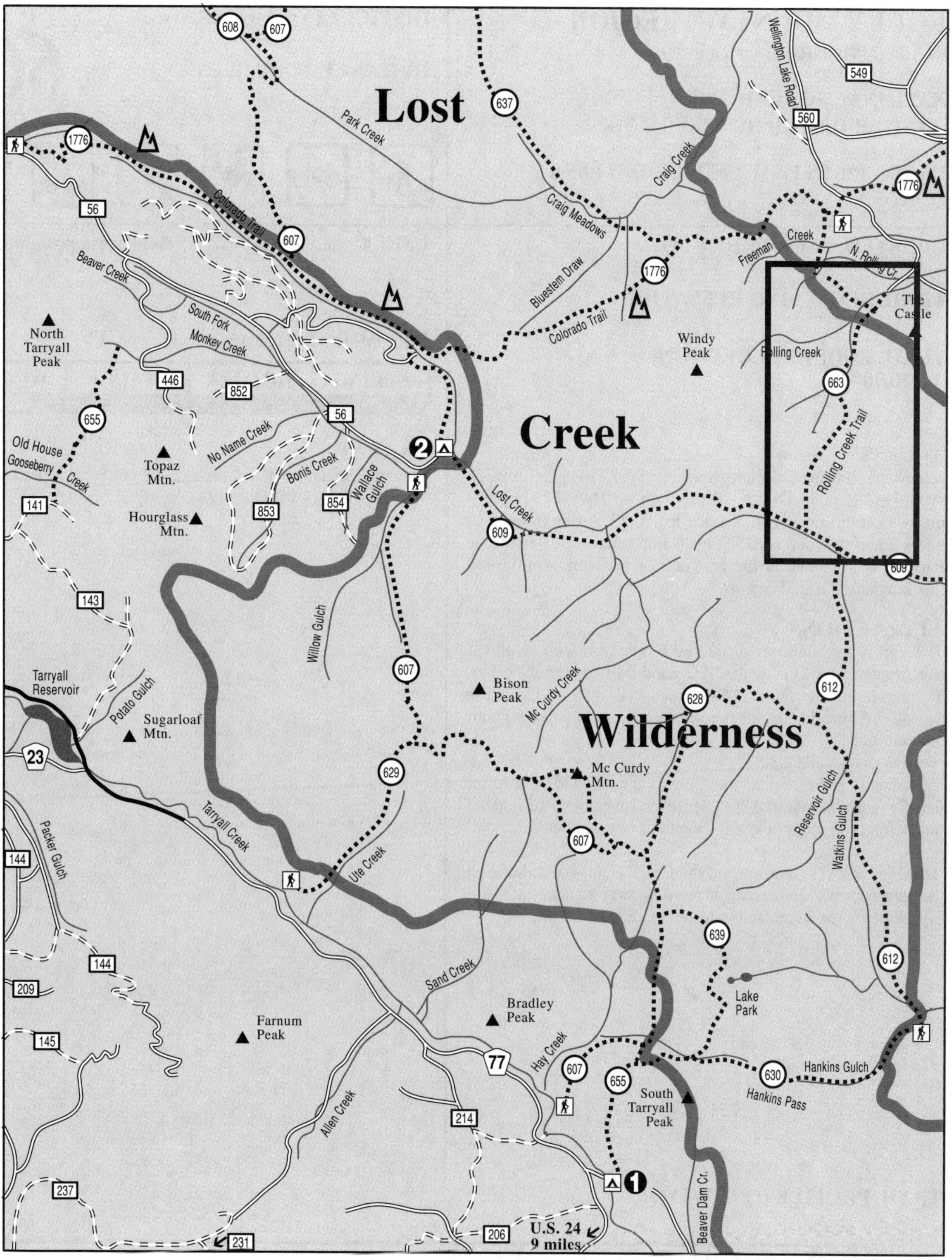

MAP 9

TRAIL #663

ROCKY MOUNTAIN REGION

MAP NUMBER(S): 16 and 19

NATIONAL FOREST: PIKE
RANGER DISTRICT: PIKES PEAK

USGS: PIKES PEAK, WOODLAND PARK QUADS

DIFFICULTY: MODERATE

DISTANCE: 2.5 MILES

TRAIL
664
CRAGS

UPDATED: NOVEMBER 1991

TRAIL BEGINNING ELEVATION:
10,100 ft.

TRAIL ENDING ELEVATION:
10,900 ft.

USE: Moderate, motorized vehicles are prohibited

RECOMMENDED SEASON:

SPRING	SUMMER	FALL	WINTER

ACCESS:

Access: 1. From Colorado Springs, take Hwy 24 through Woodland Park to Divide. Turn south on Hwy 67. Drive 4 miles, turn left on Forest Service Rd. #383. Approximately 5 miles down the road is the Crags Campground and Trailhead. Parking is available at the east end of the campground and this is where the trial begins.

ATTRACTIONS:

The trail is a moderately difficult 2.5 mile trail with an elevation change of 800 feet, this is a good trail for family hikes. Temperatures are about 15 degrees cooler than Colorado Springs. Afternoon thunderstorms are common throughout the summer.

Camping is permitted at the Crags Campground. Campfire permits are not required but please be sure your fire is dead out before you leave. Pack it in/Pack it out applies.

Drinking water is available at the Crags Campground during the summer only. Hikers are advised to carry an adequate supply. Water from streams should be treated before use.

NARRATIVE:

The trail leads to "The Crags" a group of rock pinnacle formations. This is a good cross country ski trail.

TRAIL PROFILE (ONE-WAY):

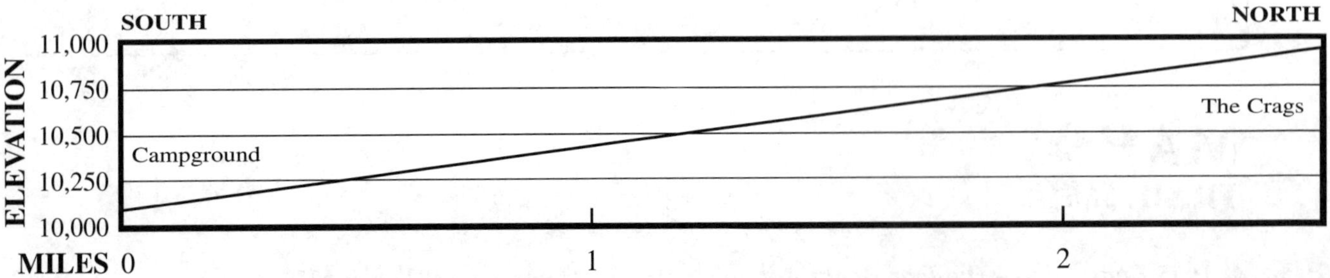

MAP 16 - South Half

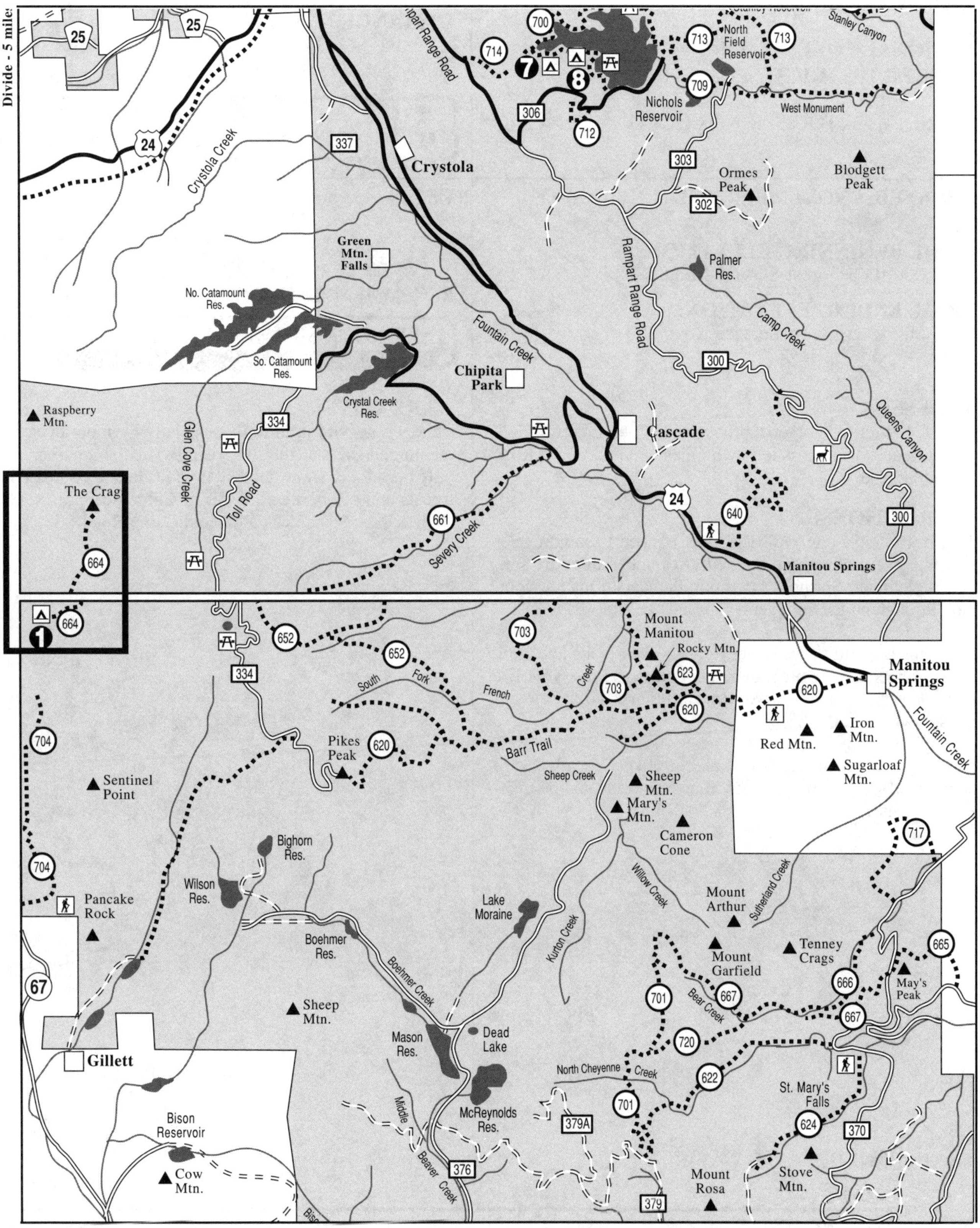

MAP 19 - North Half

TRAIL #664

ROCKY MOUNTAIN REGION

MAP NUMBER(S): 19

NATIONAL FOREST: PIKE
RANGER DISTRICT: PIKES PEAK

USGS: MANITOU SPRINGS QUAD

DIFFICULTY: MODERATE

DISTANCE: 2.5 MILES

TRAIL 665 PENROSE

UPDATED: NOVEMBER 1991

TRAIL BEGINNING ELEVATION:
7,000 ft. at Forest Service Road 370

TRAIL ENDING ELEVATION:
7,800 ft. at Forest Service Road 380

ACCESS:

Access 1. From Colorado Springs, take 26th street to Gold Camp Road. Approximately two miles up Gold Camp Road is the trailhead.

ATTRACTIONS:

The trail is moderately difficult and is open to motorcycles which are very often seen on the trail. Temperatures are similar to Colorado Springs. Afternoon thunderstorms are common throughout the summer.

Camping Is permitted in undeveloped sites along the trail. Campfire permits are not required but please be sure your fire is dead out before you leave. Pack it in/Pack it out rule applies.

Drinking water is not readily available. Hikers are advised to carry an adequate supply. Water from streams should be treated before use.

USE: Moderate

RECOMMENDED SEASON:

SPRING	SUMMER	FALL	WINTER

NARRATIVE:

This is a heavily used trail, especially by motorcycles. It is recommended that they use it only in the summer months only because of snow depths. It winds in and out of cool pine forests and offers scenic vistas of Colorado Springs, Garden Of the Gods, Pikes Peak and Gold Camp Road.

TRAIL PROFILE (ONE-WAY):

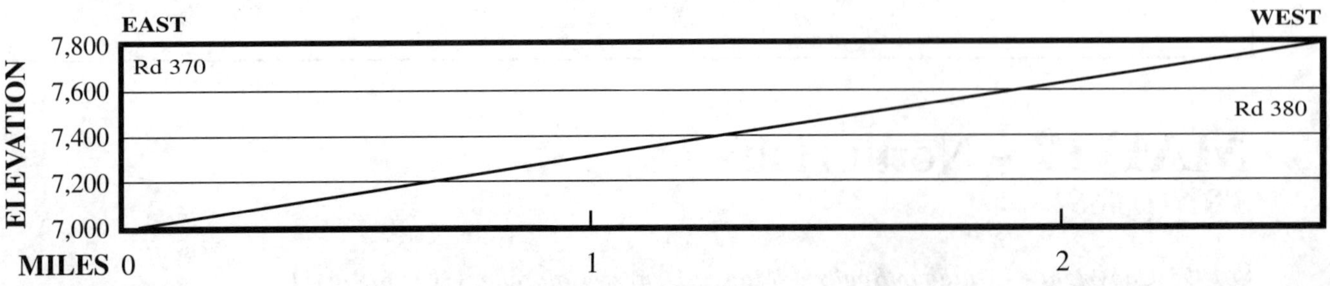

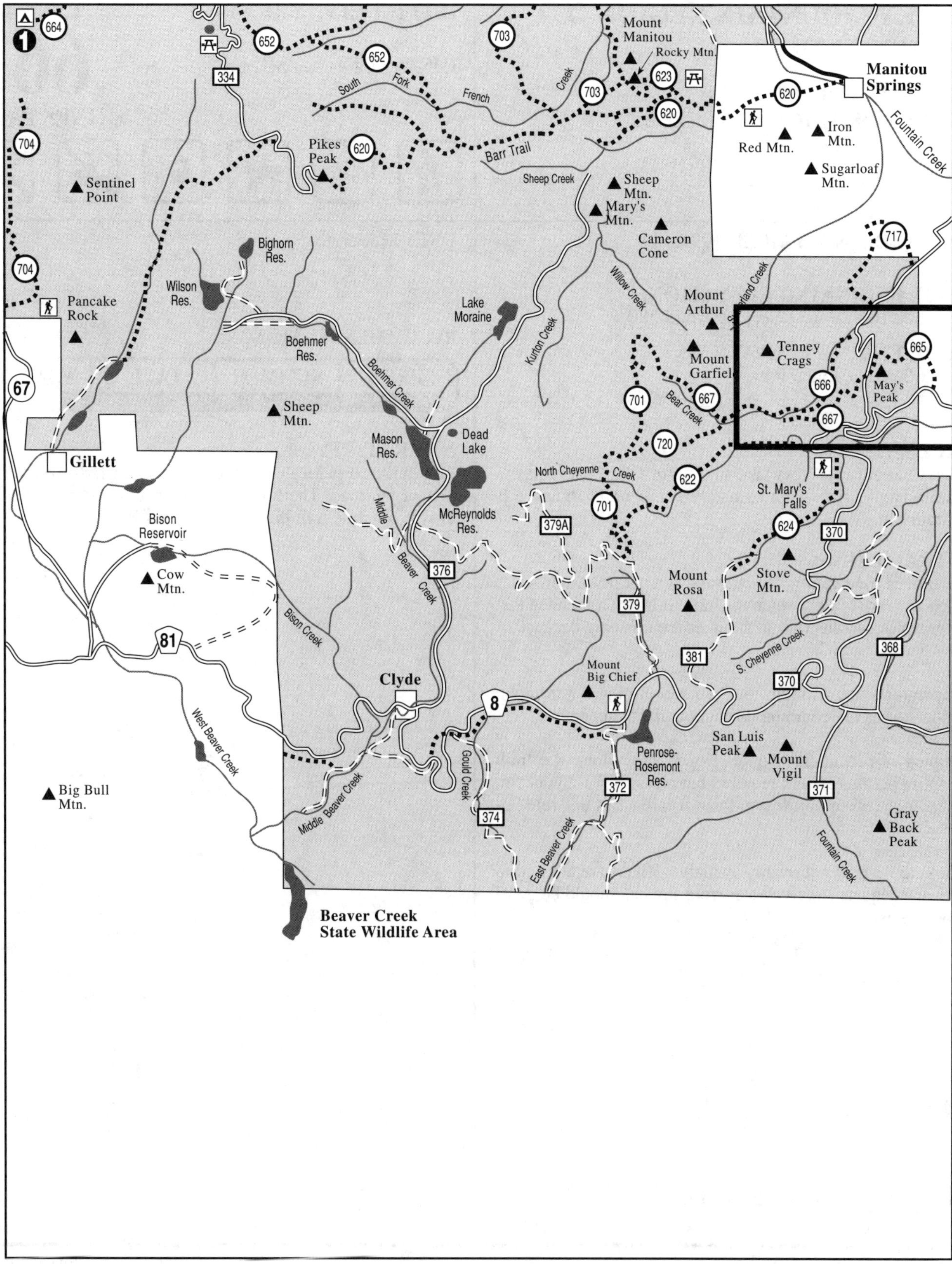

MAP 19

TRAIL #665

ROCKY MOUNTAIN REGION

MAP NUMBER(S): 19

NATIONAL FOREST: PIKE
RANGER DISTRICT: PIKES PEAK

USGS: MANITOU SPRINGS QUAD

DIFFICULTY: MODERATE

DISTANCE: 4.0 MILES

TRAIL 667 JONES PARK

UPDATED: NOVEMBER 1991

TRAIL BEGINNING ELEVATION:
7,800 ft. at Forest Service Road 380

TRAIL ENDING ELEVATION:
9,600 ft. at Jones Park

ACCESS:
From Colorado Springs, take the North Cheyenne Canyon Road to High Drive. Approximately 1 mile up High Drive Is the trailhead.

ATTRACTIONS:
The trail is moderately difficult and is open to motorcycles which are very often seen on the trail. It is recommended that motorcycles use this trail in summer months only because of snow depth.

Temperatures are similar to Colorado Springs. Afternoon thunderstorms are common throughout the summer.

Camping is permitted in undeveloped sites along the trail. Campfire permits are not required but please be sure your fire is dead out before you leave. Pack it in/Pack it out rule applies.

Drinking water is not readily available. Hikers are advised to carry an adequate supply. Water from streams should be treated before use.

USE: Moderate

RECOMMENDED SEASON:

SPRING	SUMMER	FALL	WINTER

NARRATIVE:
The trailhead is located at the top of High Drive on the left side of the road. The trail leads in a southwesterly direction. This is a 4 mile trail that leads to Jones Park.

TRAIL PROFILE (ONE-WAY):

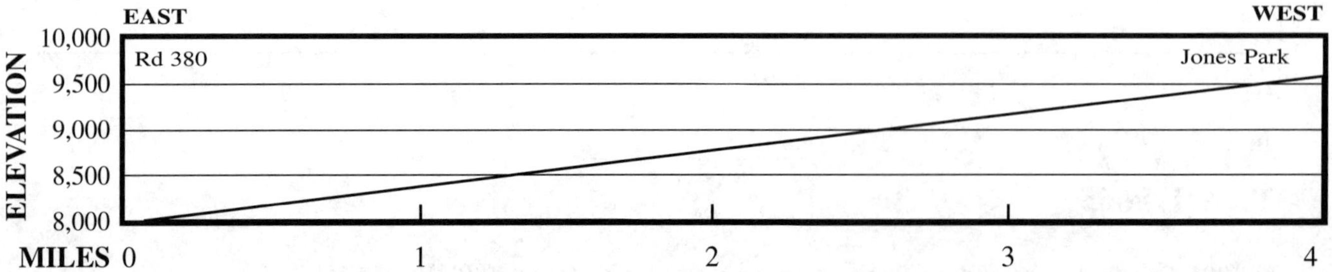

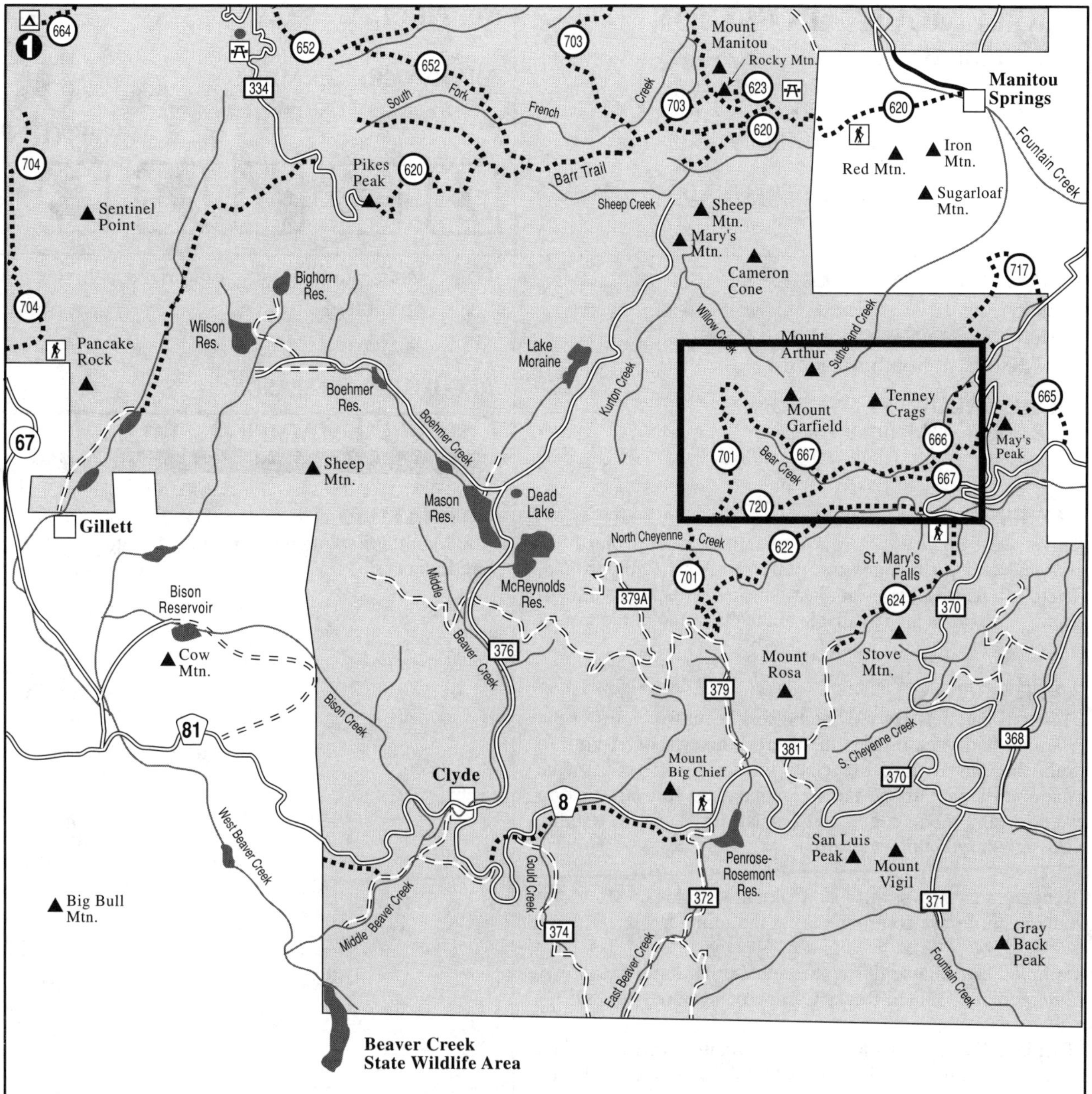

MAP 19

TRAIL #667

ROCKY MOUNTAIN REGION

MAP NUMBER(S): 16

NATIONAL FOREST: PIKE
RANGER DISTRICT: PIKES PEAK

USGS: MOUNT DECEPTION QUAD

DIFFICULTY: EASY

DISTANCE: 2.1 MILES

TRAIL 669
MANITOU PARK

UPDATED: NOVEMBER 1991

TRAIL BEGINNING ELEVATION:
7,880 ft. at South Meadows

TRAIL ENDING ELEVATION:
7,760 ft. at Manitou Lake

ACCESS:
Travel west on Hwy 24 to Woodland Park. Turn north on Hwy 67 and go 6 miles to South Meadows Campground. There is a paved parking lot just outside the campground entrance. Parking is also available at the other end of the trail at Manitou Lake Picnic Ground.

ATTRACTIONS:
This trail parallels Hwy 67 and offers beautiful views of the Manitou Park area It takes the average hiker about 1 hour to walk this paved trail. Bicyclists must yield to pedestrians. Manitou Picnic Ground has 42 picnic sites, toilets, drinking water during the summer, and a small lake open to fishing. Cross country skiing is allowed.

Temperatures are similar to Colorado Springs. Afternoon thunderstorms are common during the summer.

Camping is available during the summer at South Meadows, Colorado and Painted Rocks Campgrounds along the trail.

Drinking water is available only at Manitou Picnic Ground during the summer. Hikers are advised to carry an adequate supply.

USE: Moderate to Heavy, motorized vehicles are prohibited

RECOMMENDED SEASON:

SPRING	SUMMER	FALL	WINTER

NARRATIVE:
The Manitou Park area was a summer hunting ground for the Ute Indians.

TRAIL PROFILE (ONE-WAY):

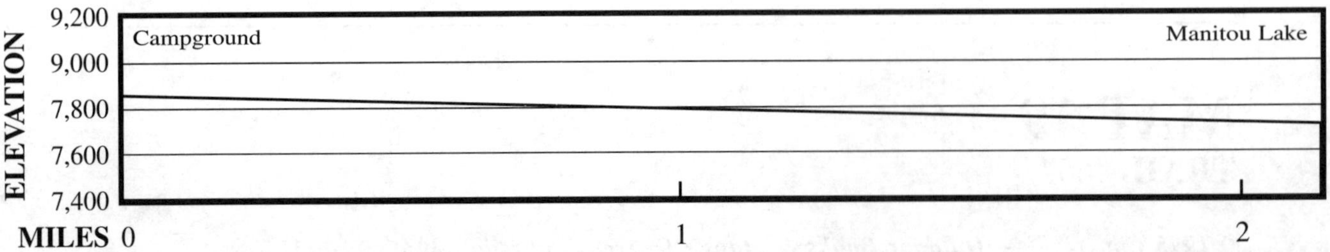

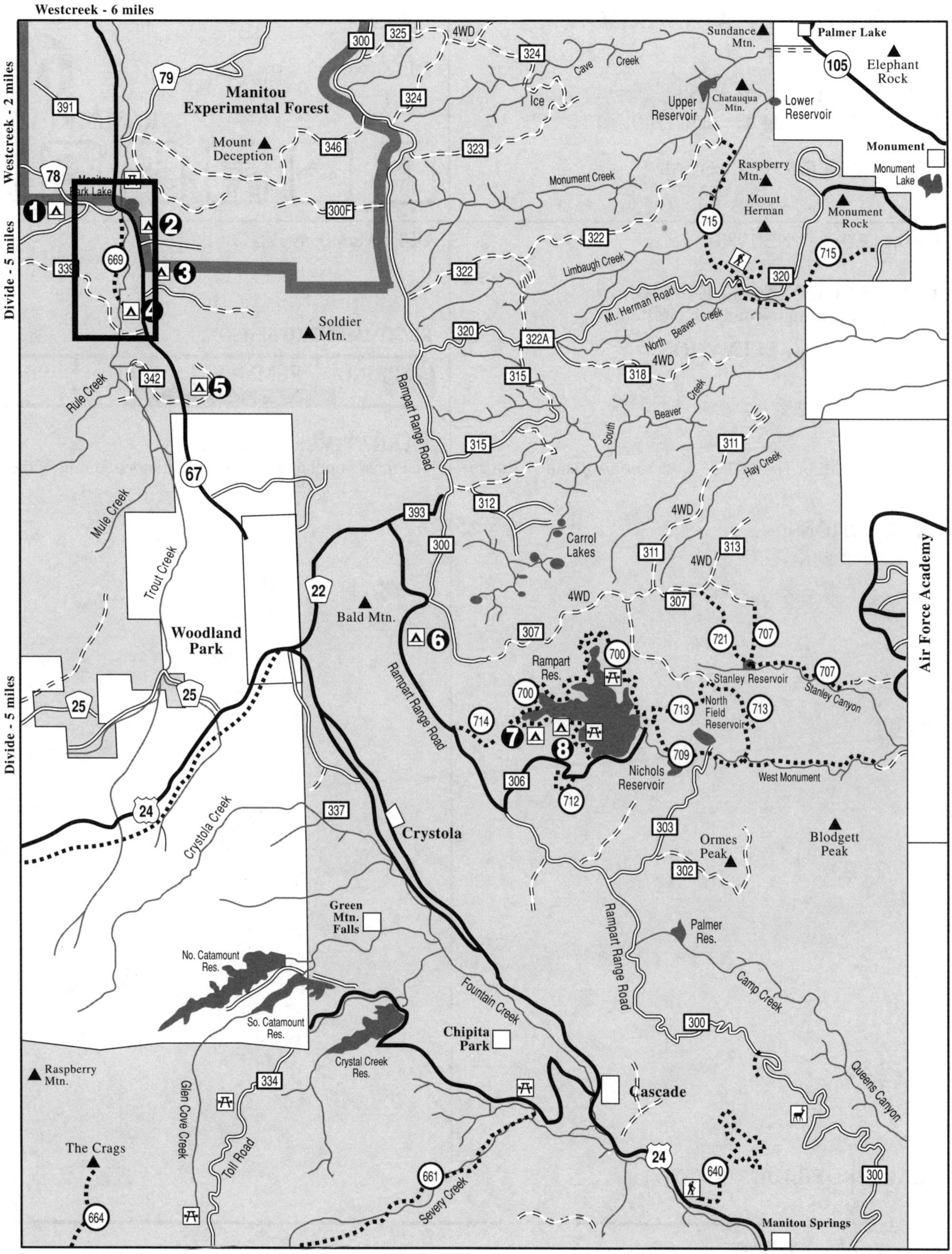

MAP 16
TRAIL #669

ROCKY MOUNTAIN REGION

MAP NUMBER(S): 11

NATIONAL FOREST: PIKE
RANGER DISTRICT: SOUTH PLATTE

USGS: DEVILS HEAD QUAD

DIFFICULTY: INTERMEDIATE

DISTANCE: 4.0 MILES

TRAIL 673
RAMPART RANGE

UPDATED: JUNE 1988

TRAIL BEGINNING ELEVATION:
Rampart Range Road - 8,256 ft.

TRAIL ENDING ELEVATION:
Sprucewood - 7,160 ft.

ACCESS:
4.2 miles south of Hwy 67 on west side of Rampart Range Road.

ATTRACTIONS:

USE: Very heavy

RECOMMENDED SEASON:

SPRING	SUMMER	FALL	WINTER

NARRATIVE:
The most popular and most used motorized trail in the area

TRAIL PROFILE (ONE-WAY):

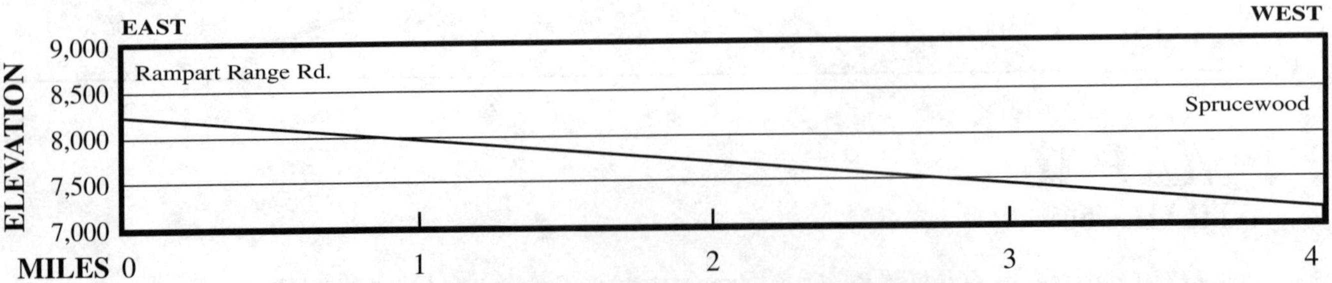

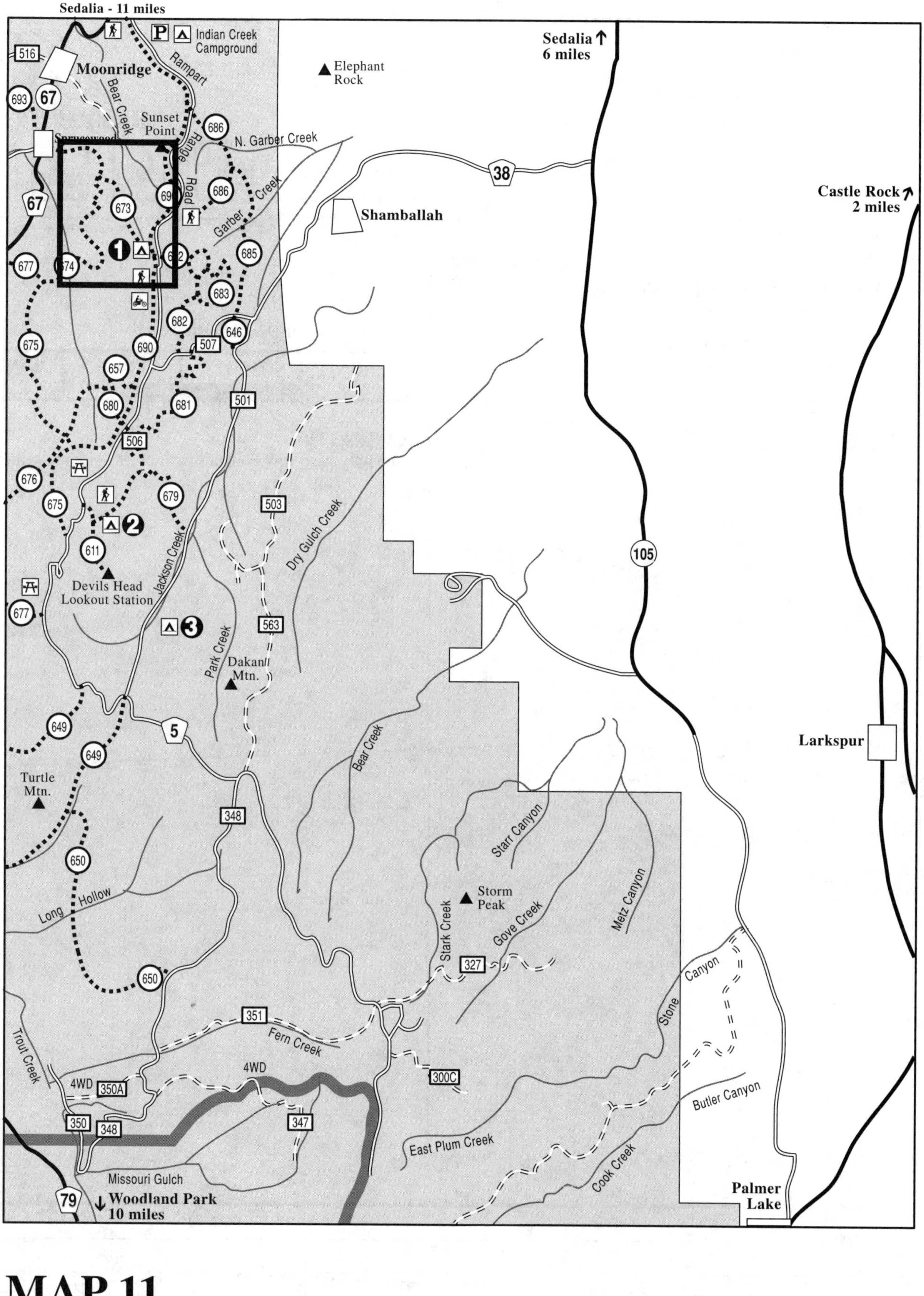

MAP 11

TRAIL #673

ROCKY MOUNTAIN REGION

MAP NUMBER(S): 11

NATIONAL FOREST: PIKE
RANGER DISTRICT: SOUTH PLATTE

USGS: DEVILS HEAD

DIFFICULTY: INTERMEDIATE

DISTANCE: 4.1 MILES

TRAIL 674

RAMPART RANGE

UPDATED: JUNE 1988

TRAIL BEGINNING ELEVATION:
7,600 ft. at Trail 673

TRAIL ENDING ELEVATION:
7,600 ft. at Trial 675

ACCESS:

ATTRACTIONS:

USE: Heavy

RECOMMENDED SEASON:

SPRING	SUMMER	FALL	WINTER

NARRATIVE:
A heavily used trail connecting 2 other heavily used trails

TRAIL PROFILE (ONE-WAY):

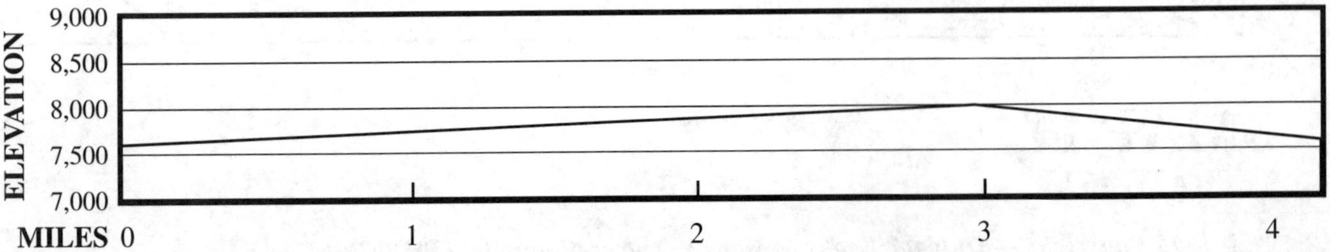

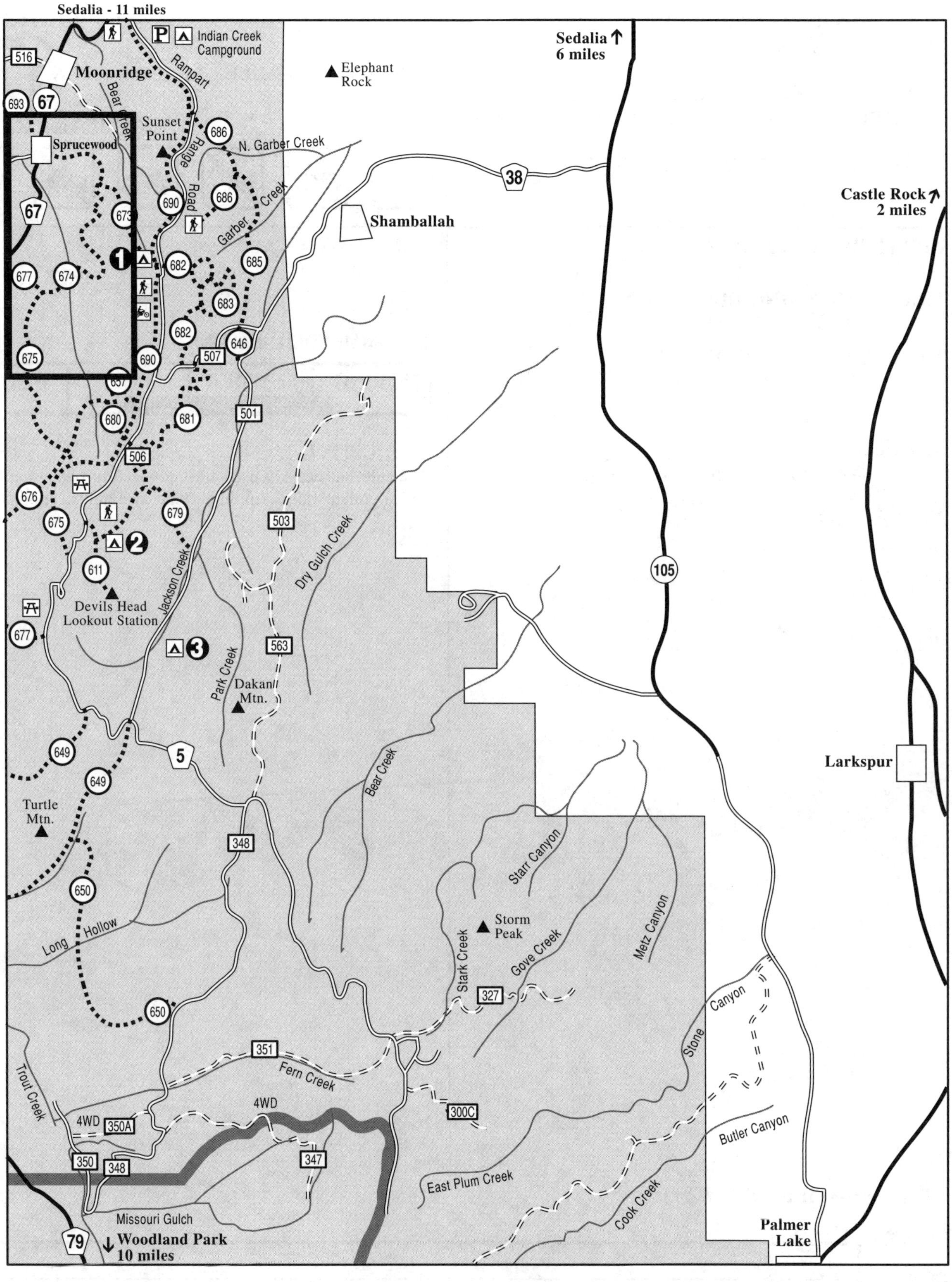

MAP 11

TRAIL #674

ROCKY MOUNTAIN REGION
MAP NUMBER(S): 11

NATIONAL FOREST: PIKE
RANGER DISTRICT: SOUTH PLATTE

USGS: DEVILS HEAD, DECKERS QUAD

DIFFICULTY: EASY

DISTANCE: 4 MILES

TRAIL 675
RAMPART RANGE

UPDATED: UNKNOWN

TRAIL BEGINNING ELEVATION:
Rampart Road - 8,780 ft.

TRAIL ENDING ELEVATION:
Sugar Creek Road - 7,960 ft.

ACCESS:
8 miles south of Hwy 67 on Rampart Range Road

ATTRACTIONS:

USE: Heavy.

RECOMMENDED SEASON:

SPRING	SUMMER	FALL	WINTER

NARRATIVE:
A scenic but heavily used trail connecting two major roads and six other trails. Ample parking available at trailhead

TRAIL PROFILE (ONE-WAY):

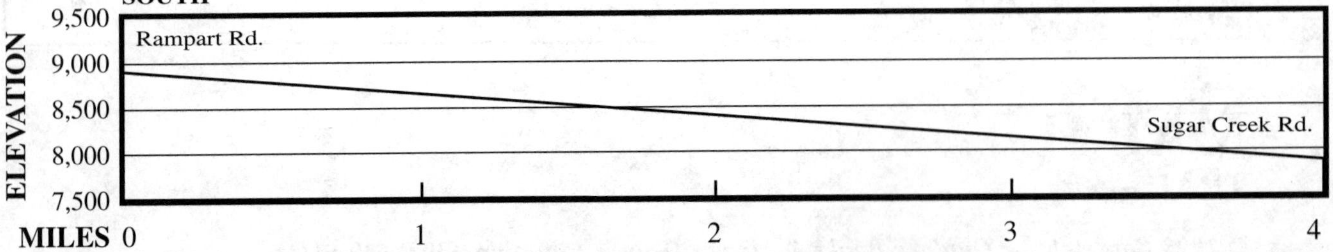

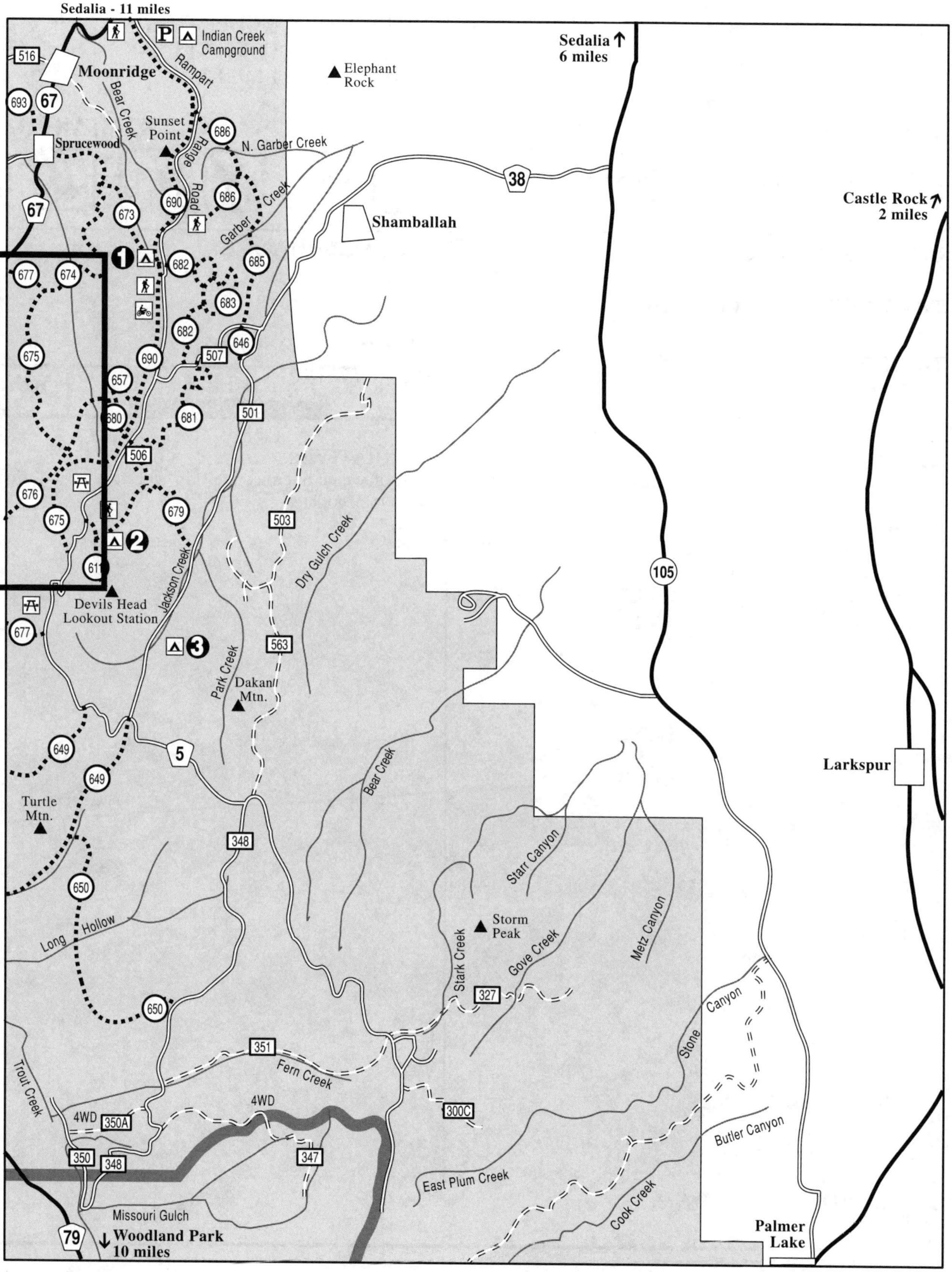

MAP 11

TRAIL #675

ROCKY MOUNTAIN REGION

MAP NUMBER(S): 10 and 11

NATIONAL FOREST: PIKE
RANGER DISTRICT: SOUTH PLATTE

USGS: DEVILS HEAD, DECKERS QUADS

DIFFICULTY: MODERATE TO DIFFICULT
DISTANCE: 6.1 MILES

TRAIL 676

RAMPART RANGE

UPDATED: UNKNOWN

TRAIL BEGINNING ELEVATION:
8,400 ft. at trail # 675

TRAIL ENDING ELEVATION:
7,800 ft. at trail # 677

ACCESS:
From trailhead of 675 proceed 0.8 miles to trail 676

ATTRACTIONS:

USE: Moderate

RECOMMENDED SEASON:

SPRING	SUMMER	FALL	WINTER

NARRATIVE:
A very scenic intermediate trail with two very difficult sections to negotiate.

TRAIL PROFILE (ONE-WAY):

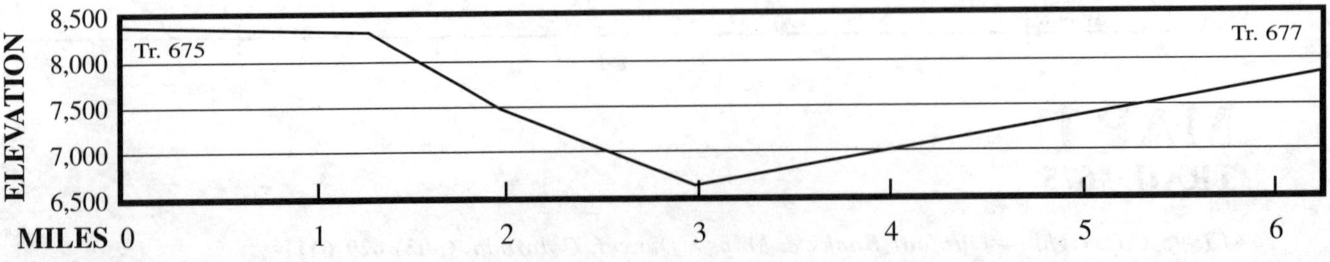

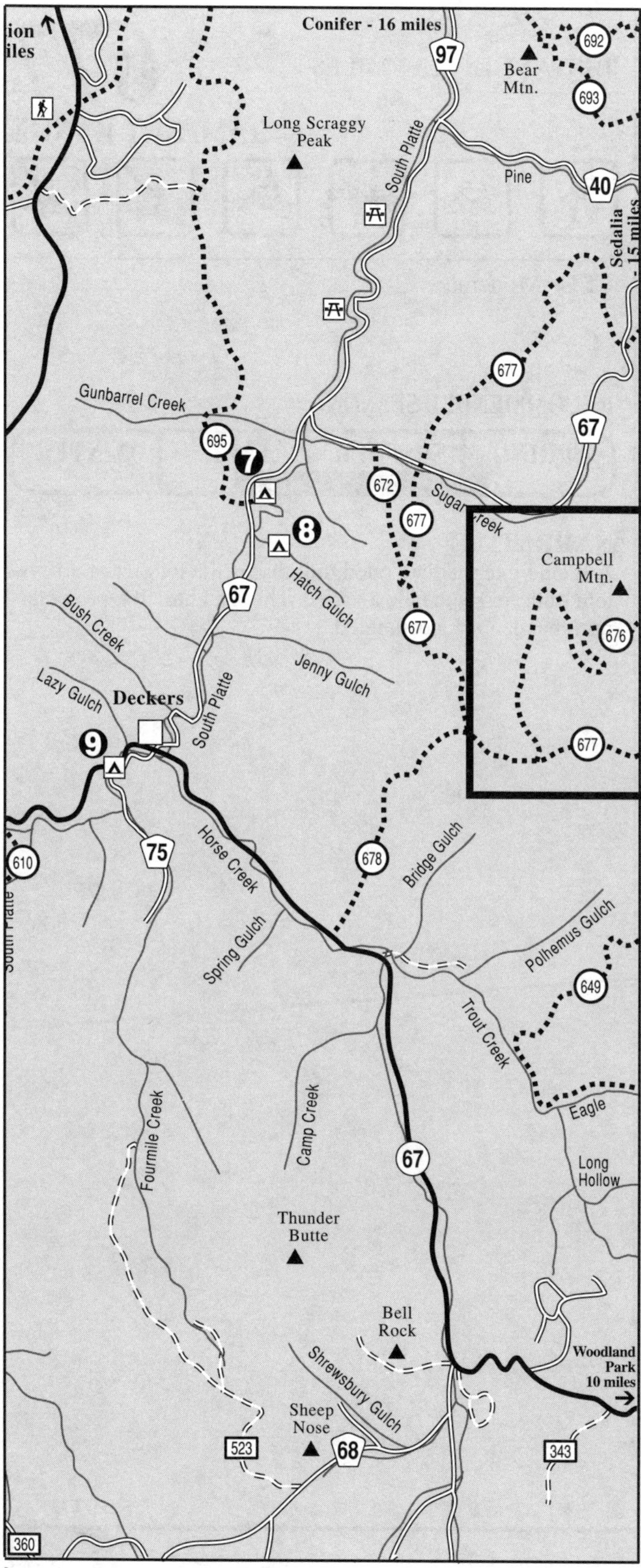

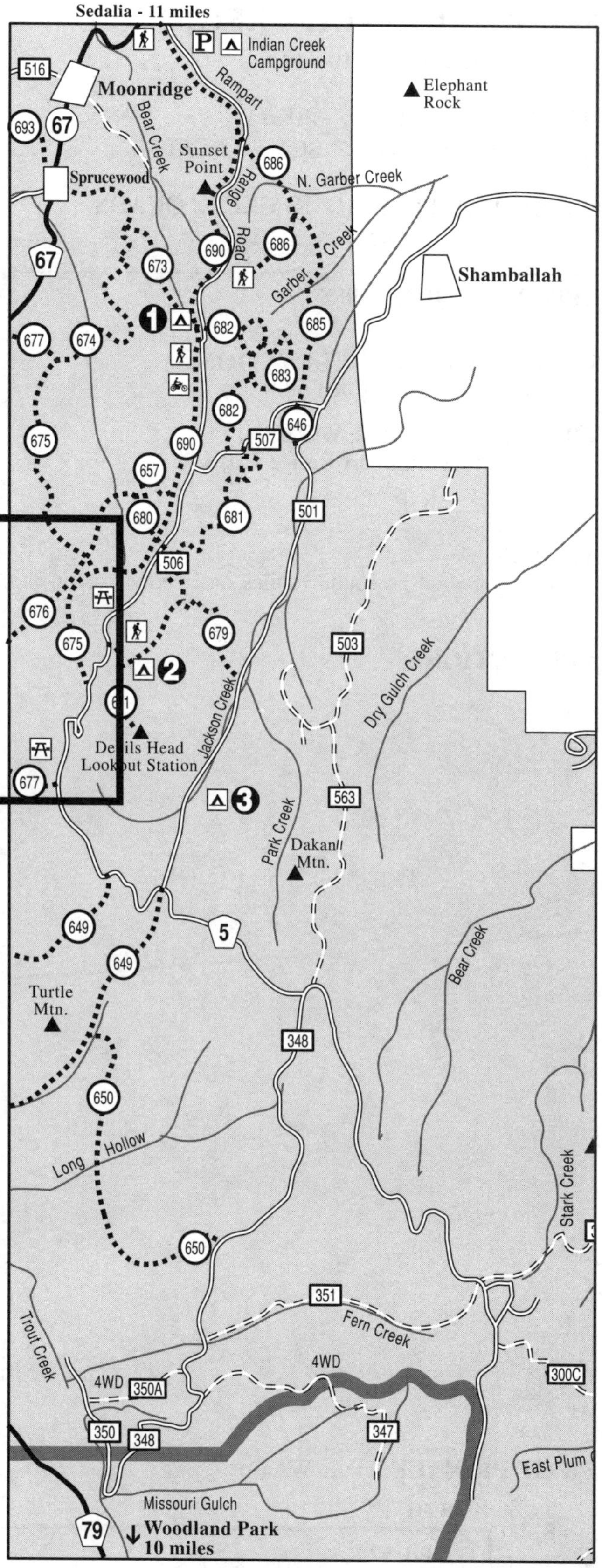

MAP 10 - East Half

TRAIL #676

MAP 11 - West Half

ROCKY MOUNTAIN REGION

MAP NUMBER(S): 10 and 11

NATIONAL FOREST: PIKE
RANGER DISTRICT: SOUTH PLATTE

USGS: DEVILS HEAD, DECKERS QUADS

DIFFICULTY: INTERMEDIATE

DISTANCE: 12.3 MILES

TRAIL 677
RAMPART RANGE

UPDATED: JUNE 1988

TRAIL BEGINNING ELEVATION:
7,920 ft. at County Rd 67

TRAIL ENDING ELEVATION:
8,757 ft. at Rampart Range Road

ACCESS:
From Sprucewood go south 3 miles on County Road 67 to trailhead

ATTRACTIONS:

USE: Moderate

RECOMMENDED SEASON:

SPRING | SUMMER | FALL | WINTER

NARRATIVE:
A scenic intermediate rated trail that is the longest of all system trails and is the most varied. This trail has all types of terrain that can be encountered

TRAIL PROFILE (ONE-WAY):

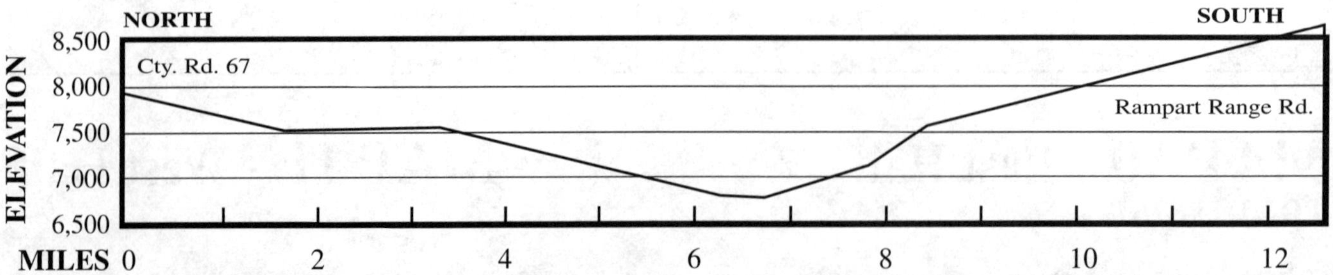

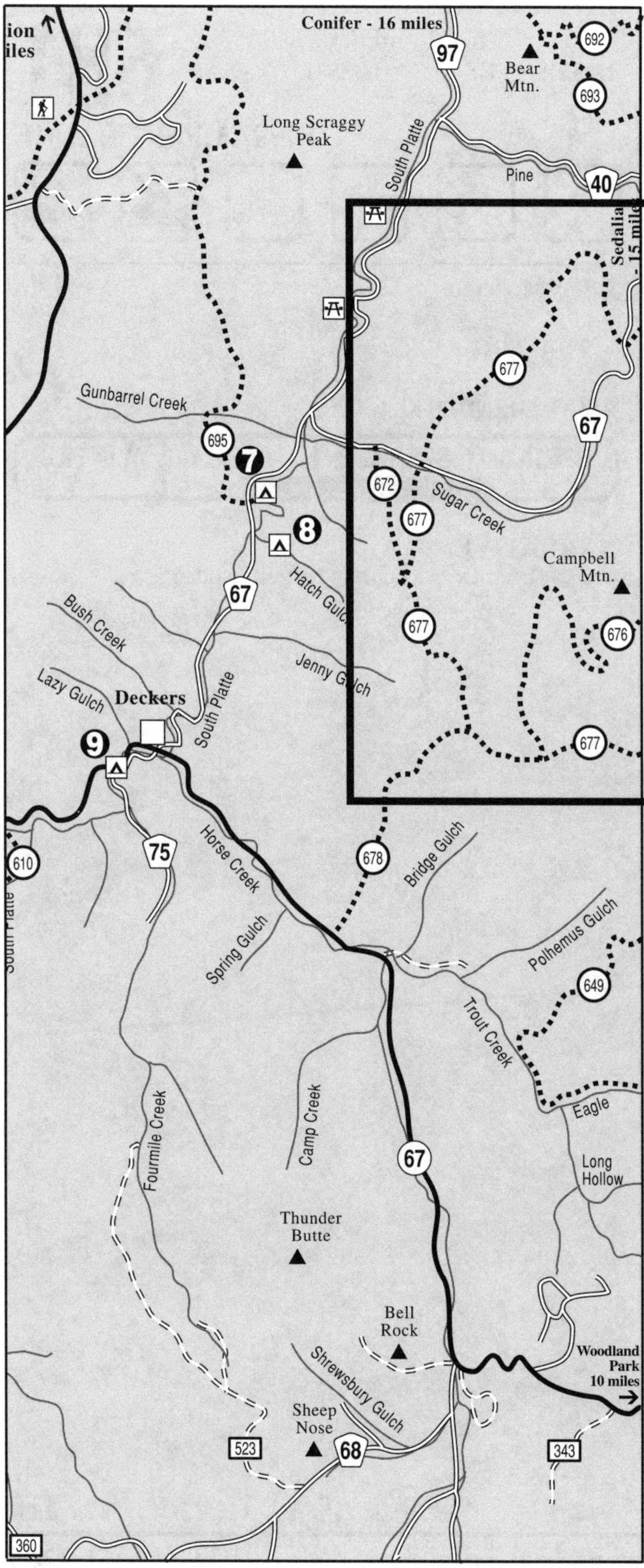

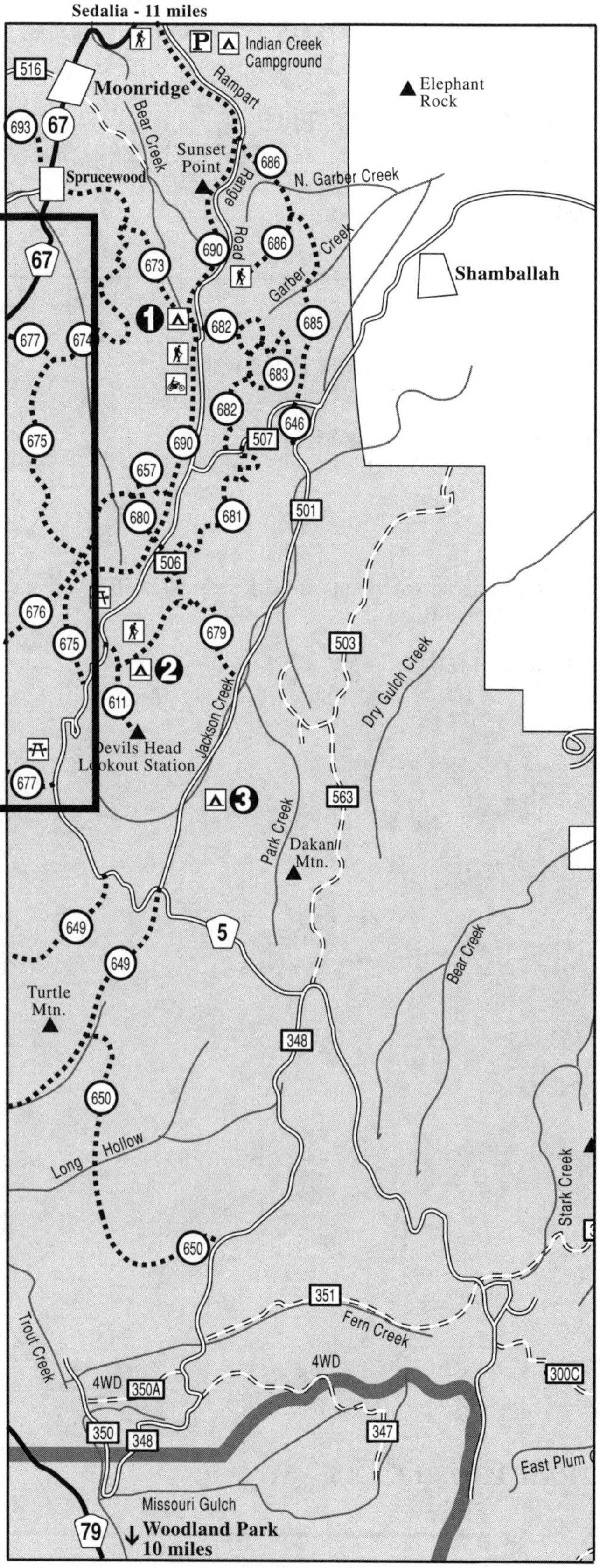

MAP 10 - East Half

TRAIL #677

MAP 11 - West Half

ROCKY MOUNTAIN REGION

MAP NUMBER(S): 11

NATIONAL FOREST: PIKE
RANGER DISTRICT: SOUTH PLATTE

USGS:

DIFFICULTY: MODERATE TO DIFFICULT
DISTANCE: 5.8 MILES

TRAIL 681

RAMPART RANGE

UPDATED: JUNE 1988

TRAIL BEGINNING ELEVATION:
8,080 ft. at Jackson Creek Rd. (502)

TRAIL ENDING ELEVATION:
8,400 ft. at Trail 679

ACCESS:
This trail has access points from Roads 502, 507, and the Rampart Range Road.

ATTRACTIONS:
Near Devils Head Campground

USE: Moderate

RECOMMENDED SEASON:

SPRING	SUMMER	FALL	WINTER

NARRATIVE:
Trail 681 serves as a connecting trail to other areas.

TRAIL PROFILE (ONE-WAY):

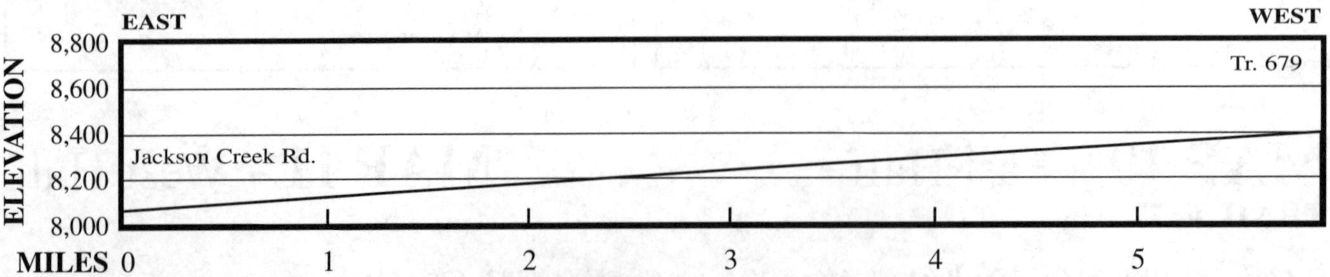

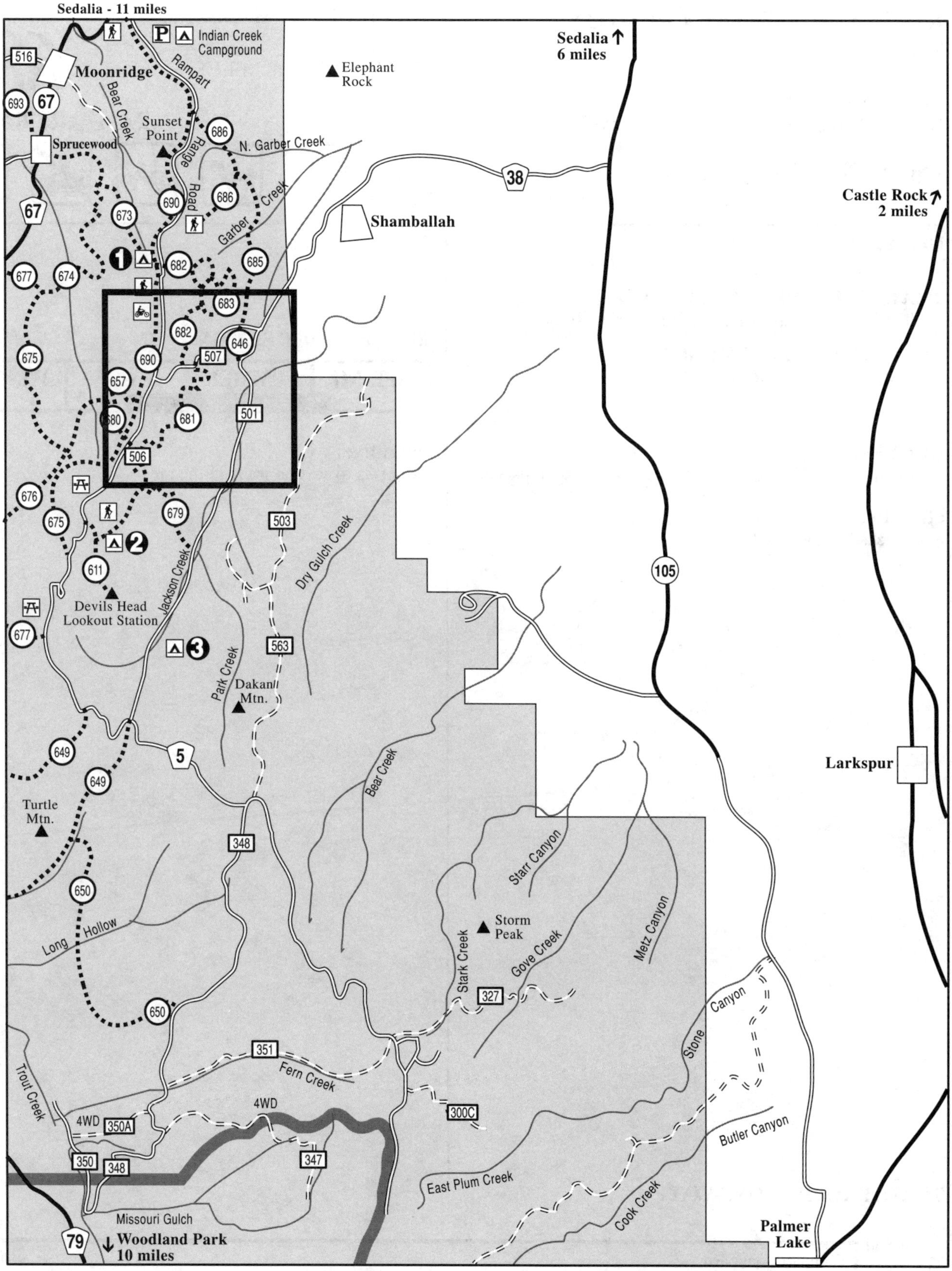

MAP 11

TRAIL #681

ROCKY MOUNTAIN REGION

MAP NUMBER(S): 11

NATIONAL FOREST: PIKE
RANGER DISTRICT: SOUTH PLATTE

USGS: DEVILS HEAD QUAD

DIFFICULTY: MODERATE

DISTANCE: 2.8 MILES

TRAIL 682
RAMPART RANGE

UPDATED: JUNE 1988

TRAIL BEGINNING ELEVATION:
8,120 ft., Rampart Range Rd. at Flat Rocks

TRAIL ENDING ELEVATION:
8,450 ft. at Rim Road #502

ACCESS:
South on Rampart Range Road 4 miles to Flat Rock parking.

attractions:
Flat Rocks Campground

USE: Moderate

RECOMMENDED SEASON:

SPRING	SUMMER	FALL	WINTER

narrative:
Connects to several trails in the area.

TRAIL PROFILE (ONE-WAY):

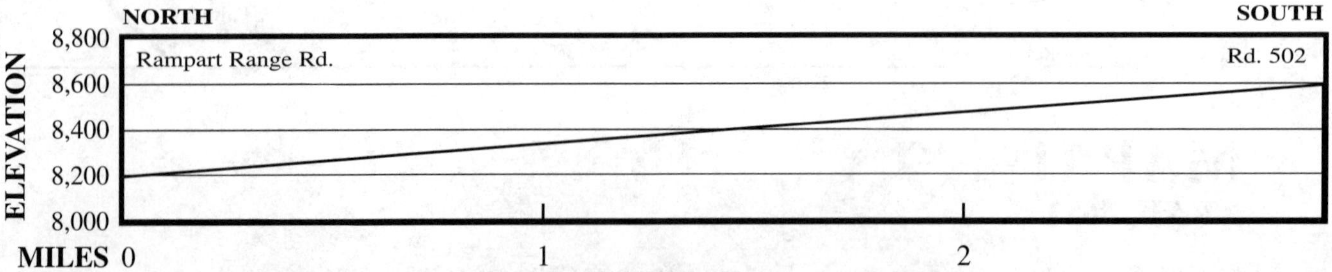

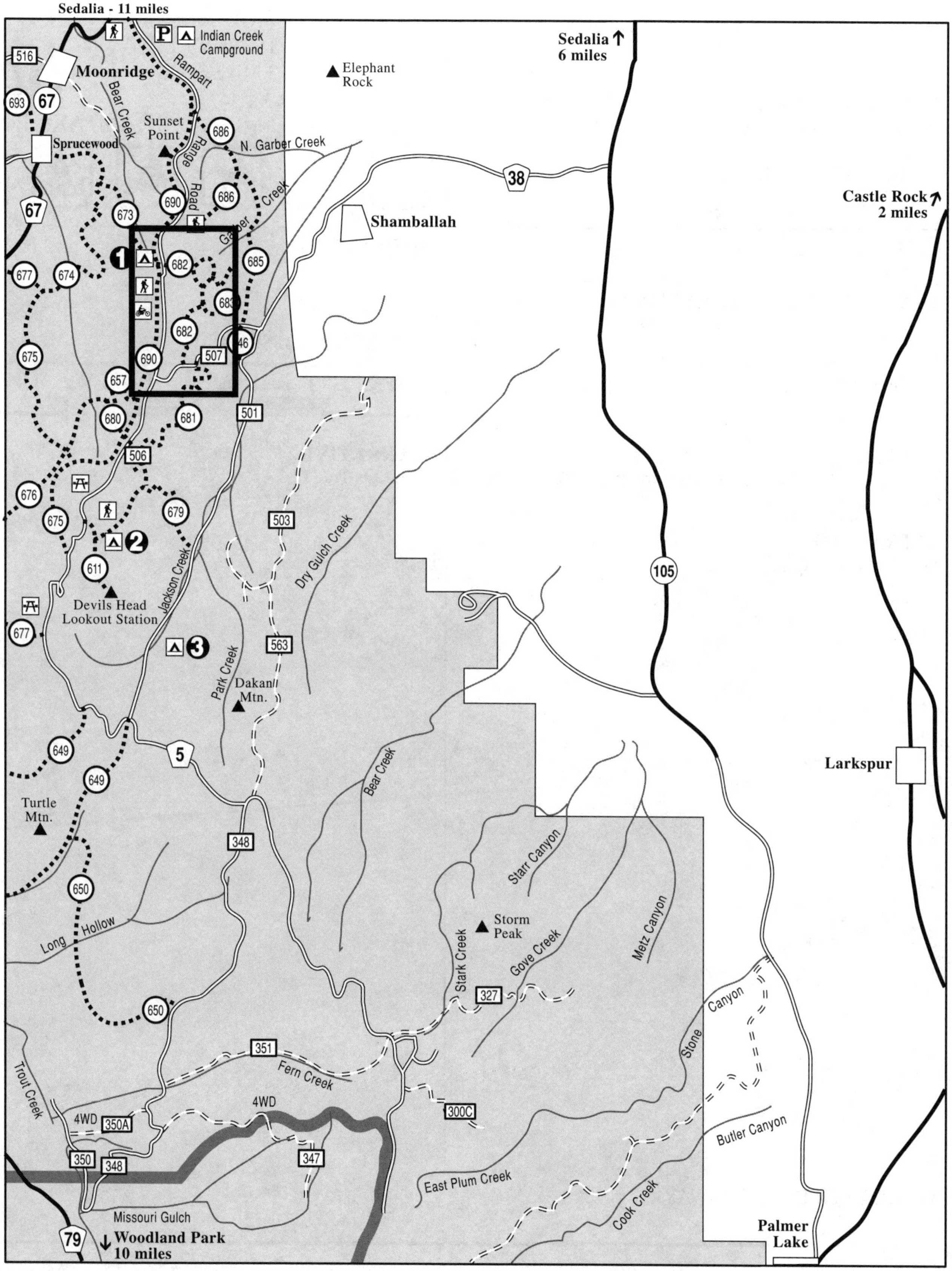

MAP 11

TRAIL #682

ROCKY MOUNTAIN REGION

MAP NUMBER(S): 11

NATIONAL FOREST: PIKE
RANGER DISTRICT: SOUTH PLATTE

USGS: DEVILS HEAD QUAD

DIFFICULTY: MODERATE TO DIFFICULT
DISTANCE: 3.2 MILES

TRAIL 685
RAMPART RANGE

UPDATED: JUNE 1988

USE: Moderate

TRAIL BEGINNING ELEVATION:
From Rim Road #502 - 8,000 ft.

TRAIL ENDING ELEVATION:
Junction of trail #686 - 7,580 ft.

RECOMMENDED SEASON:

SPRING	SUMMER	FALL	WINTER

ACCESS:
Trailhead is 2 miles west of Rampart Range Road on Rim Road 507

ATTRACTIONS:
Near Flat Rocks Campground

NARRATIVE:
Connects Trail 686

TRAIL PROFILE (ONE-WAY):

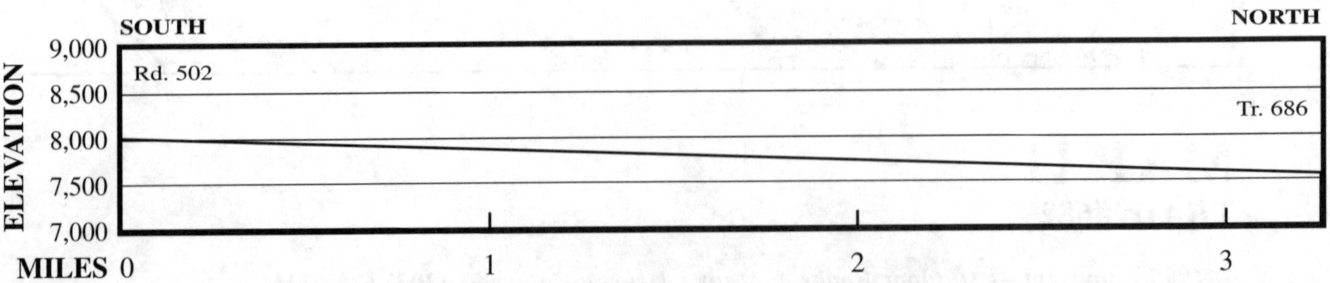

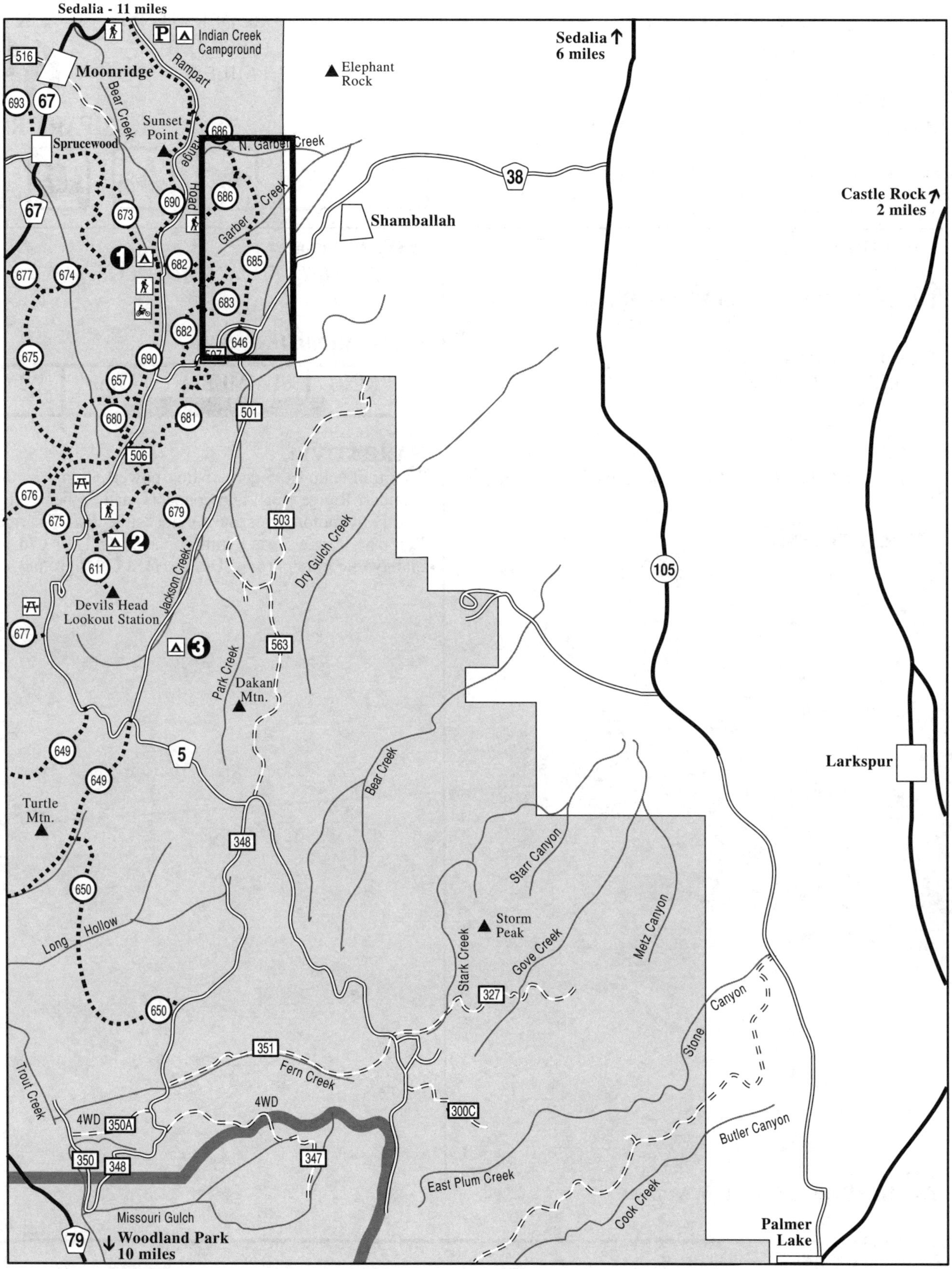

MAP 11

TRAIL #685

ROCKY MOUNTAIN REGION
MAP NUMBER(S): 11

NATIONAL FOREST: PIKE
RANGER DISTRICT: SOUTH PLATTE

USGS: DEVILS HEAD QUAD

DIFFICULTY: MODERATE TO DIFFICULT
DISTANCE: 7.7 MILES

TRAIL 690
RAMPART RANGE

UPDATED: JUNE 1988

TRAIL BEGINNING ELEVATION:
7,600 ft. at Indian Creek Campground

TRAIL ENDING ELEVATION:
8,400 ft. at Trail 675

ACCESS:
Trailhead at Indian Creek Campground on Rampart Range Road

ATTRACTIONS:

USE: Moderate
HIKING, BIKING, MOTORCYCLING, ATV'S

RECOMMENDED SEASON:

SPRING	SUMMER	FALL	WINTER

NARRATIVE:
This trail follows under existing powerline that parallels the Rampart Range Road and connects many other trails. Used mostly by motorcycles that are not "street legal" to ride on the Rampart Range Road. Trail connects to Trail 675 approximately one mile north of Devils Head Campground.

TRAIL PROFILE (ONE-WAY):

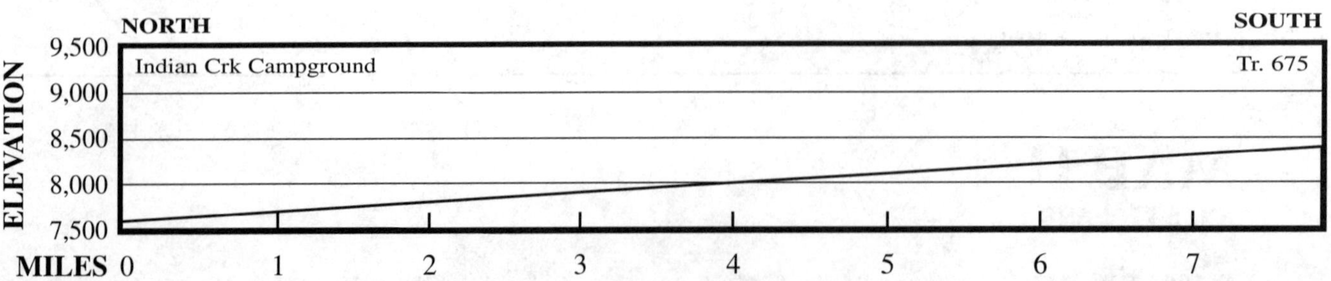

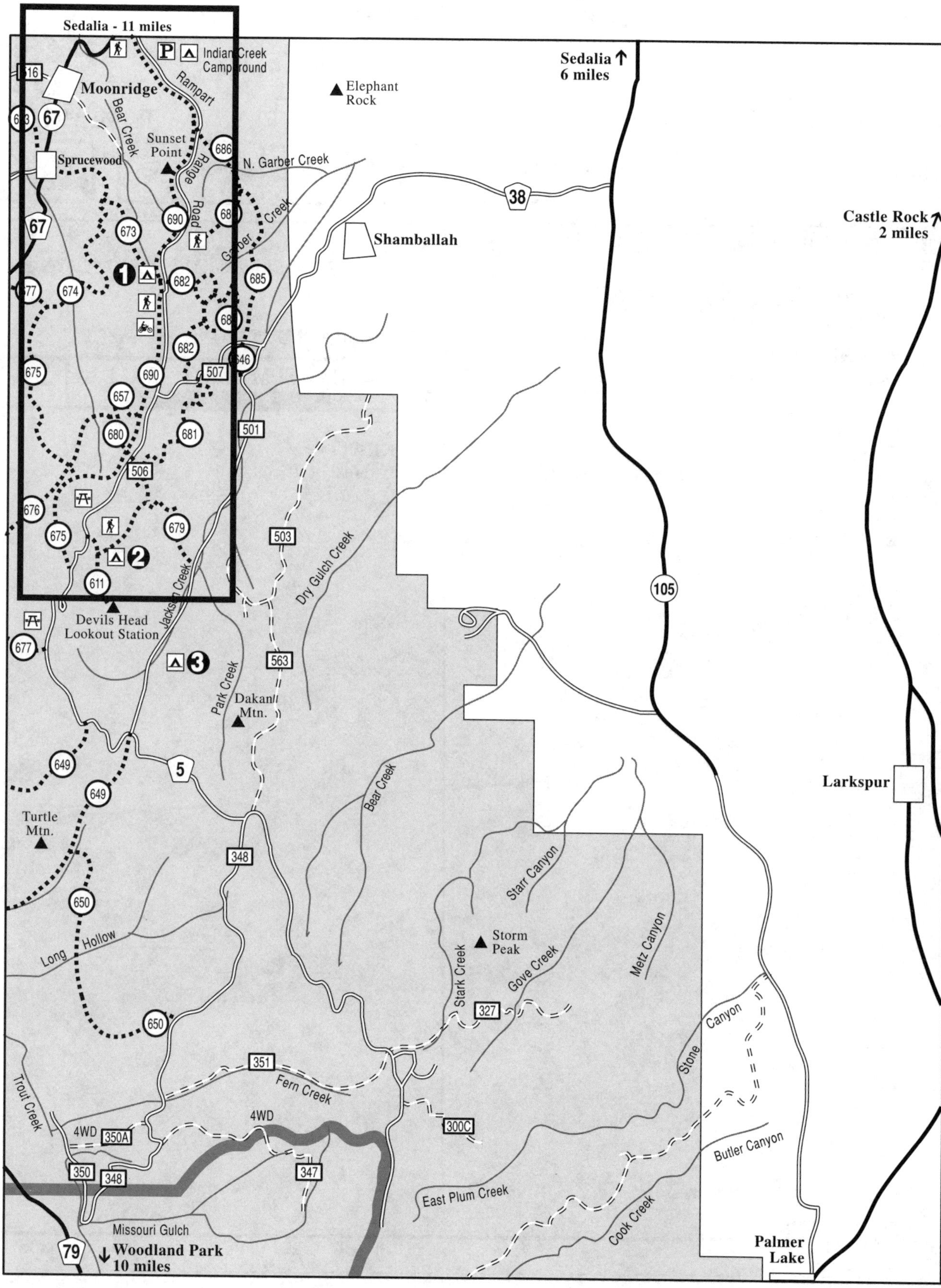

MAP 11

TRAIL #690

ROCKY MOUNTAIN REGION

MAP NUMBER(S): 4

NATIONAL FOREST: PIKE
RANGER DISTRICT: MOUNT EVANS

USGS: DECKERS, PLATTE CANYON QUADS

DIFFICULTY: DIFFICULT

DISTANCE: 3.4 MILES

TRAIL 692

RAMPART RANGE

UPDATED: JUNE 1988

TRAIL BEGINNING ELEVATION:
7,200 ft. at Trail 693

TRAIL ENDING ELEVATION:
6,800 ft. at Colorado Trail (Tr. 1776)

ACCESS:
Access from Trail 693 - two miles east of Colo. Hwy. 97 on Trail 693.

ATTRACTIONS:

USE: Low
HIKING, BIKING, MOTORCYCLING, ATV'S

RECOMMENDED SEASON:

SPRING	SUMMER	FALL	WINTER

NARRATIVE:
An extremely difficult trail

TRAIL PROFILE (ONE-WAY):

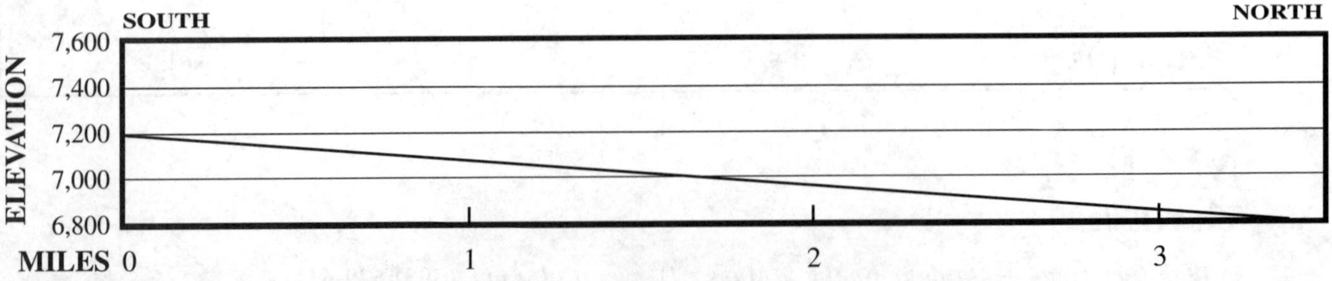

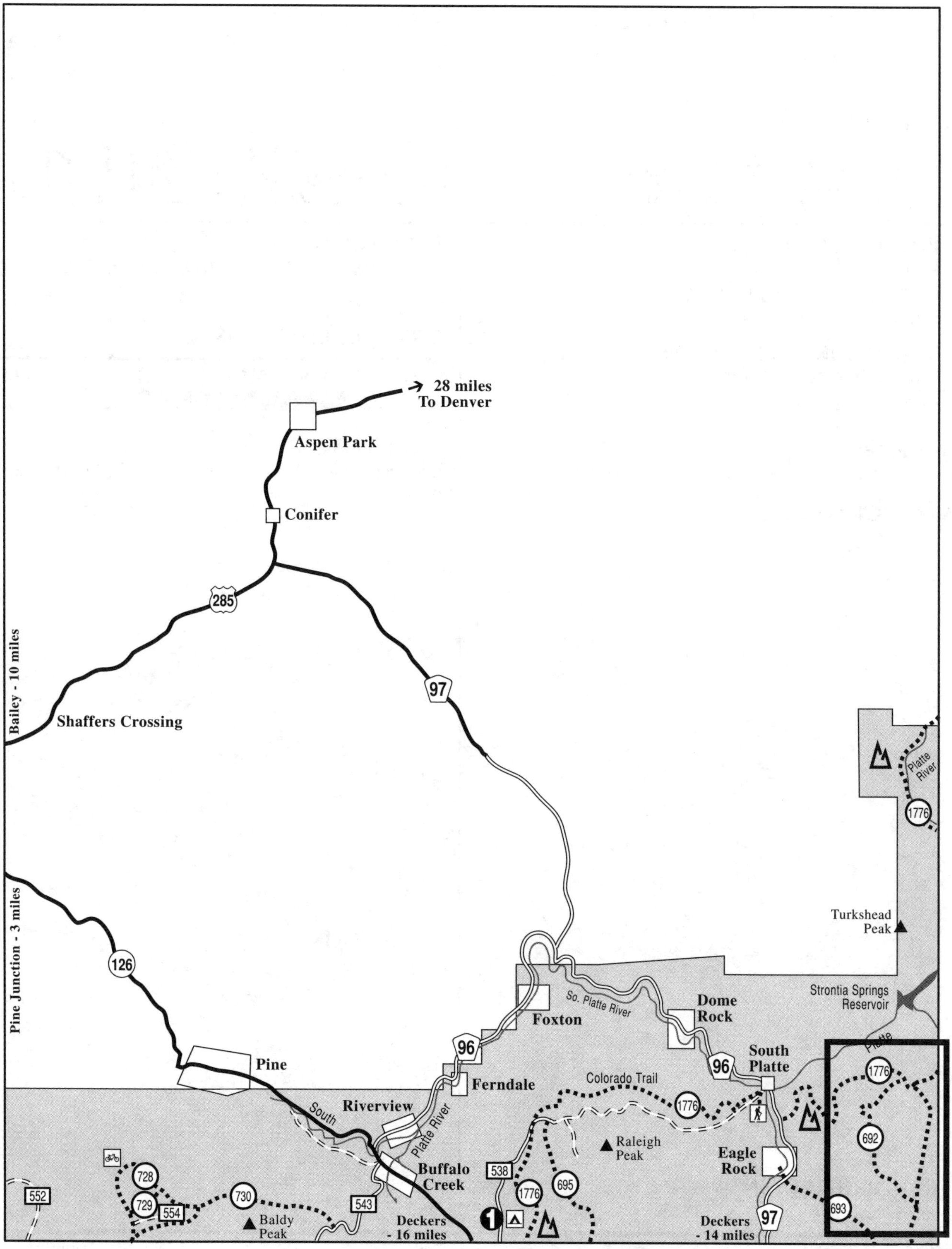

MAP 4

TRAIL #692

ROCKY MOUNTAIN REGION

MAP NUMBER(S): 4, 10 and 11

NATIONAL FOREST: PIKE
RANGER DISTRICT: SOUTH PLATTE

USGS: DEVILS HEAD, DECKERS, PLATTE CANYON QUADS

DIFFICULTY: MODERATE TO DIFFICULT
DISTANCE: 6.5 MILES

TRAIL 693
RAMPART RANGE

UPDATED: JUNE 1988

TRAIL BEGINNING ELEVATION:
6,800 ft. at Sprucewood

TRAIL ENDING ELEVATION:
6,500 ft. at Colo. Hwy. 97

ACCESS:
Parking near Sprucewood west of Colo. Hwy. 67

ATTRACTIONS:

USE: Moderate
HIKING, BIKING, MOTORCYCLING, ATV'S

RECOMMENDED SEASON:

SPRING	SUMMER	FALL	WINTER

NARRATIVE:
Intersects Trail 692 approximately 3 miles west of Sprucewood.

TRAIL PROFILE (ONE-WAY):

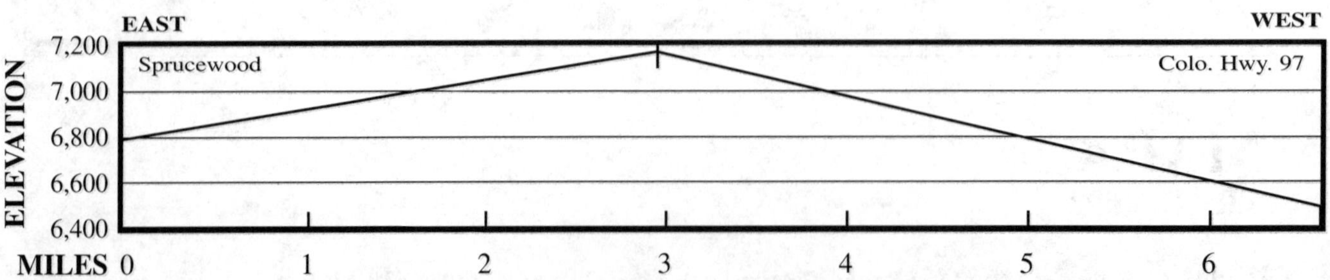

MAP 4 - Southeast Quarter

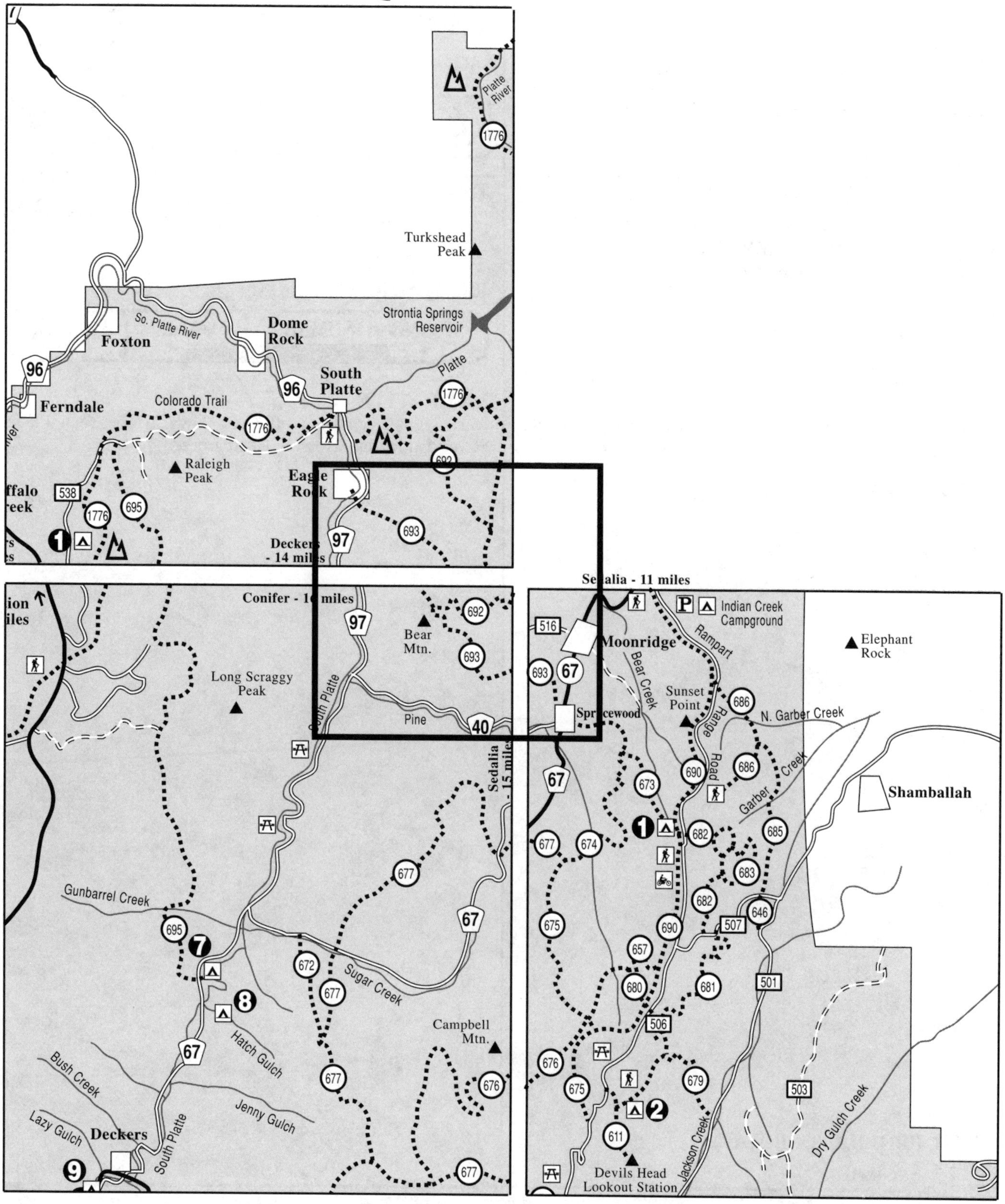

MAP 10 - Northeast Quarter

TRAIL #693

MAP 11 - Northwest Quarter

ROCKY MOUNTAIN REGION
MAP NUMBER(S): 4 and 10

NATIONAL FOREST: PIKE
RANGER DISTRICT: SOUTH PLATTE

USGS: DECKERS, PLATTE CANYON QUADS

DIFFICULTY: MODERATE

DISTANCE: 8.6 MILES

TRAIL 695
RAMPART RANGE

UPDATED: JUNE 1988

TRAIL BEGINNING ELEVATION:
6,400 ft. at Colo. Hwy. 67

TRAIL ENDING ELEVATION:
7,600 ft. at Colo. Trail #1776

ACCESS:
Trail up from South Platte townsite or trailhead near Platte River Picnic Ground.

ATTRACTIONS:

USE: Moderate
HIKING, BIKING, MOTORCYCLING, ATV'S

RECOMMENDED SEASON:

SPRING	SUMMER	FALL	WINTER

NARRATIVE:
Very scenic. Trailhead off Colo. Hwy. 67 between Platte River and Bridge Crossing Campgrounds.

TRAIL PROFILE (ONE-WAY):

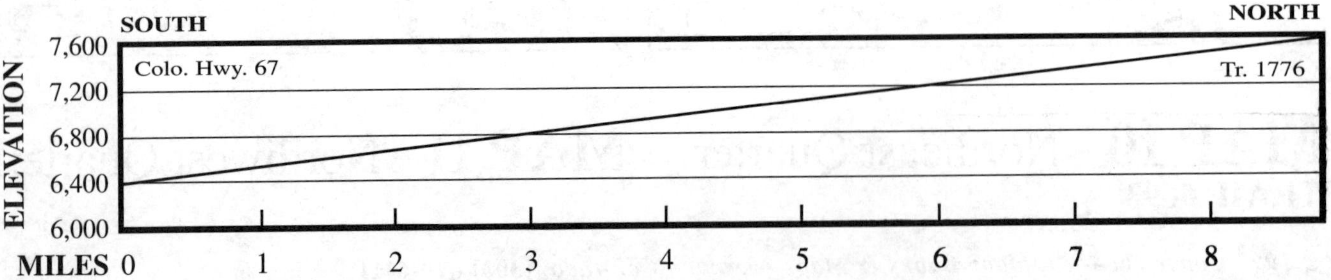

MAP 4 - South Half

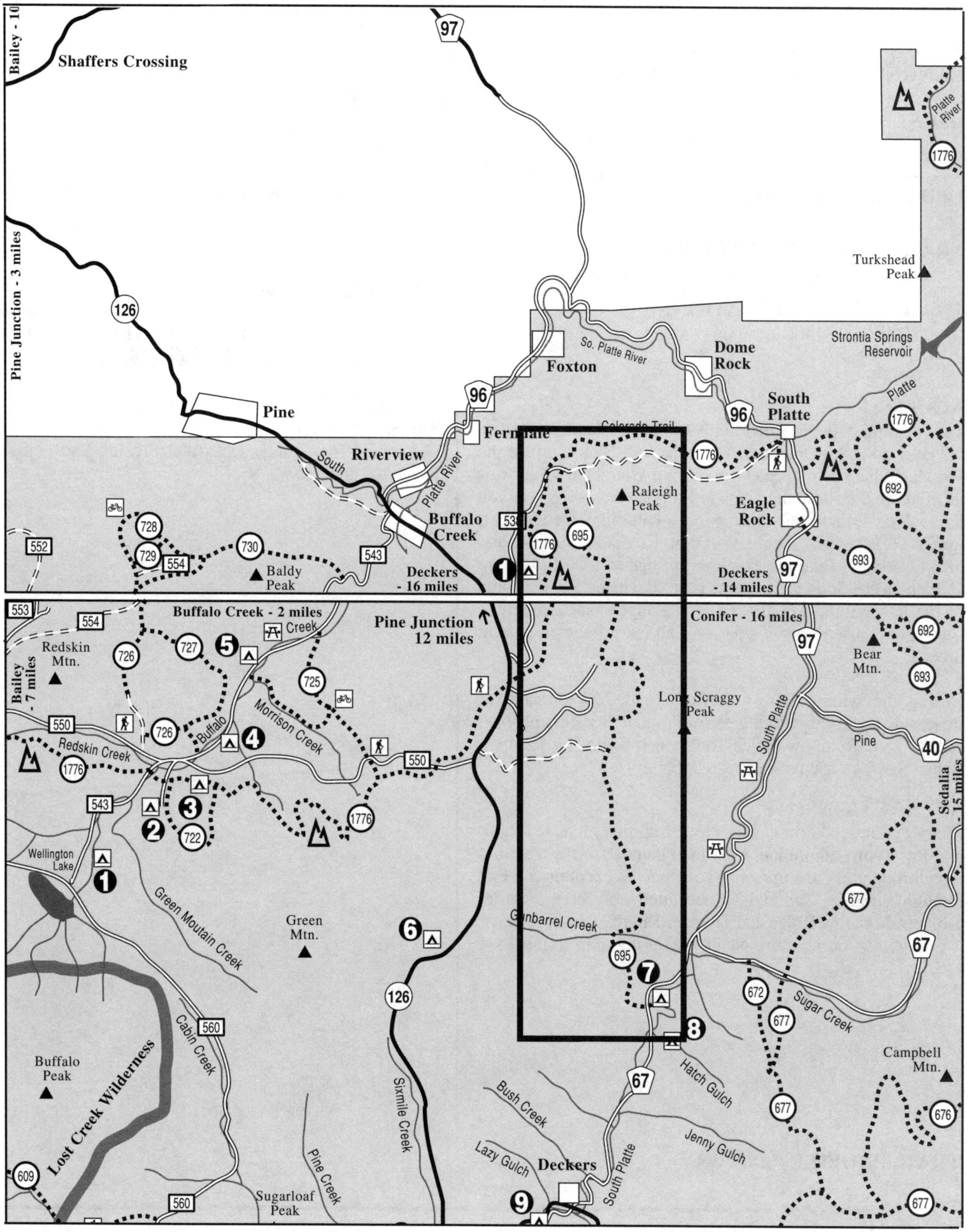

MAP 10 - North Half

TRAIL #695

ROCKY MOUNTAIN REGION

MAP NUMBER(S): 16

NATIONAL FOREST: PIKE
RANGER DISTRICT: PIKES PEAK

USGS: CASCADE QUAD

DIFFICULTY: EASY

DISTANCE: 11.6 MILES

TRAIL 700

RAMPART RESERVOIR

UPDATED: NOVEMBER 1991

USE: Moderate, motortized vehicle prohibited

TRAIL BEGINNING ELEVATION:
9,160 ft. in parking area

TRAIL ENDING ELEVATION:
9,160 ft. in parking area

RECOMMENDED SEASON:

SPRING | SUMMER | FALL | WINTER

ACCESS:
Take Rampart Range Road north from Woodland Park (turn off Hwy 24 at McDonalds). This road will take you past the Woodland Park schools and a portion of Woodland's Park residential area. Follow the signs to Rampart Reservoir. If you are looking for a scenic drive prior to your hike, the Rampart Range Road offers scenic vistas of Colorado Springs and many canyons. Take the Manitou Springs exit as if going to Garden of the Gods. After going past "Balanced Rock" a sign on the left will indicate the Rampart Range Road. This route should be chosen only if you have an extra hour of driving time.

During the winter, Forest Service Road 306 to Rampart Reservoir is closed with a gate. Additional cross country ski access is via Rainbow Gulch Trail which begins on Rampart Range Road a little west of road 306.

ATTRACTIONS:
A cross country ski trail. It is a loop trail that traverses Rampart Reservoir's shoreline. Afternoon thunderstorms are common throughout the summer. Snow may be encountered Fall through Spring. Camping is permitted only at the campgrounds, Meadow Ridge and Thunder Ridge. Drinking water is available at promontory picnic are during the summer.

NARRATIVE:
This easy trail follows the Reservoir's shoreline providing opportunities for fishing access, picnicking and enjoying the beauty of the area.

TRAIL PROFILE (ONE-WAY):

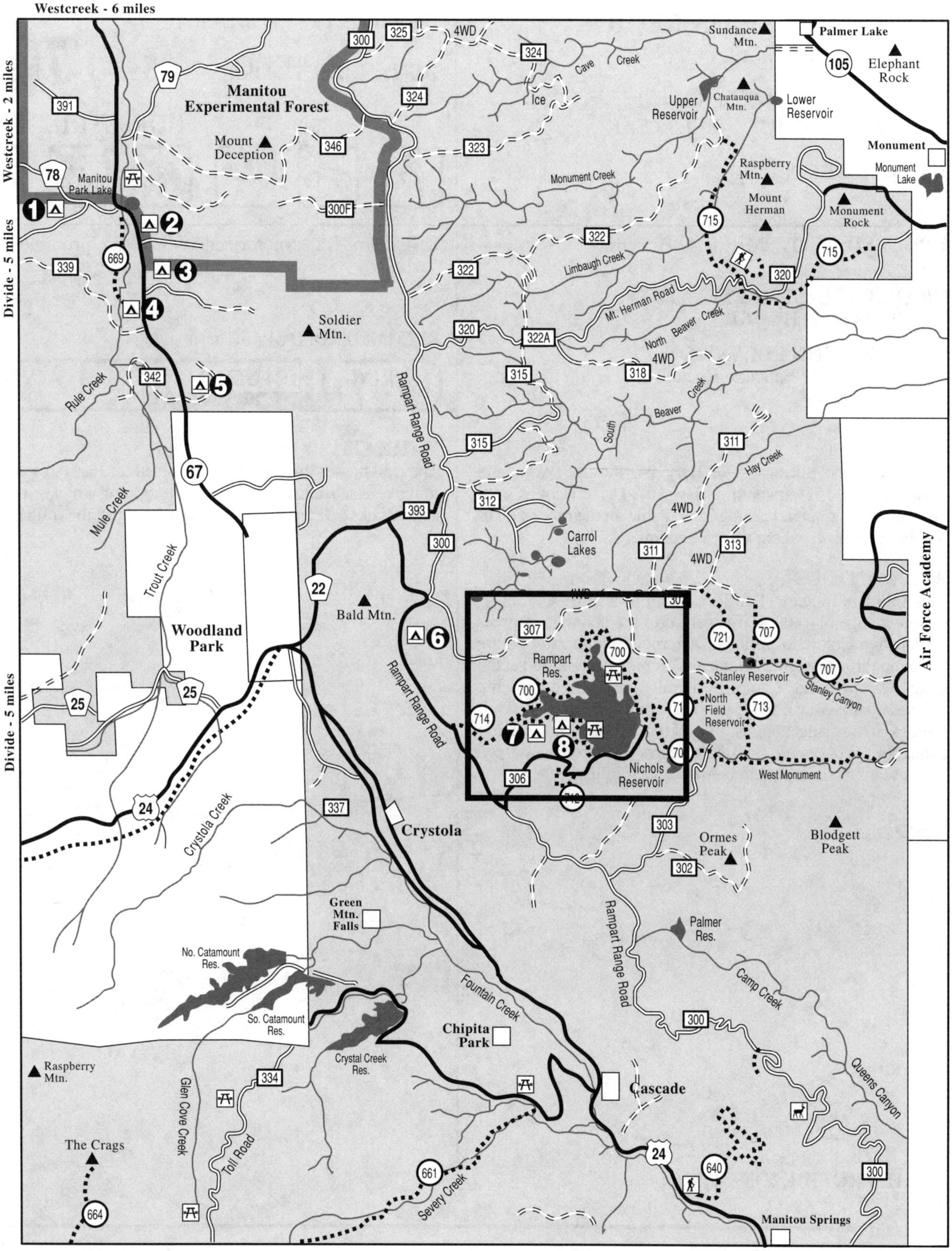

MAP 16

TRAIL #700

ROCKY MOUNTAIN REGION

MAP NUMBER(S): 19

NATIONAL FOREST: PIKE
RANGER DISTRICT: PIKES PEAK

USGS: PIKES PEAK

DIFFICULTY: MODERATE

DISTANCE: 3.0 MILES

TRAIL 704
HORSETHIEF PARK

UPDATED: NOVEMBER 1991

USE: Moderate, motorized vehicles are prohibited

TRAIL BEGINNING ELEVATION:
10,100 ft. at Hwy 67

TRAIL ENDING ELEVATION:
11,500 ft. at "Pancake Rocks"

RECOMMENDED SEASON:

SPRING	SUMMER	FALL	WINTER

ACCESS:
From Colorado Springs, take Hwy 24, through Woodland Park, to Divide. Turn south on Hwy 67. Drive 8 miles until you come to a tunnel. Parking is available on the other side of the tunnel. This is where the trail begins.

ATTRACTIONS:
This trail is a moderately difficult trail of 4 miles and a elevation change of 1400 ft. Temperatures are about 15 degrees cooler than Colorado Springs. Afternoon thunderstorms are common throughout the summer. Camping is permitted at the Crags Campground. Campfire permits are not required but please be sure your fire is dead out before you leave. Pack it in/Pack it out rule applies. Drinking water is not readily available. Hikers are advised to carry an adequate supply. Water from streams should be treated before use.

NARRATIVE:
The trail passes through the old Horsethief Park area and culminates at an interesting rock outcropping known as "Pancake Rocks". From there you must return to the trailhead.

TRAIL PROFILE (ONE-WAY):

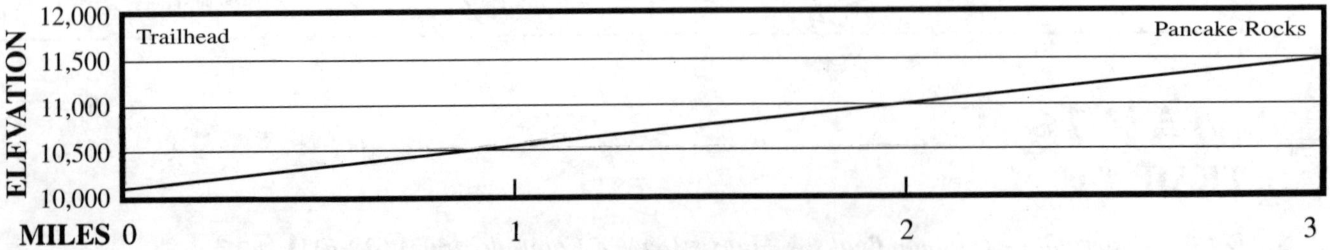

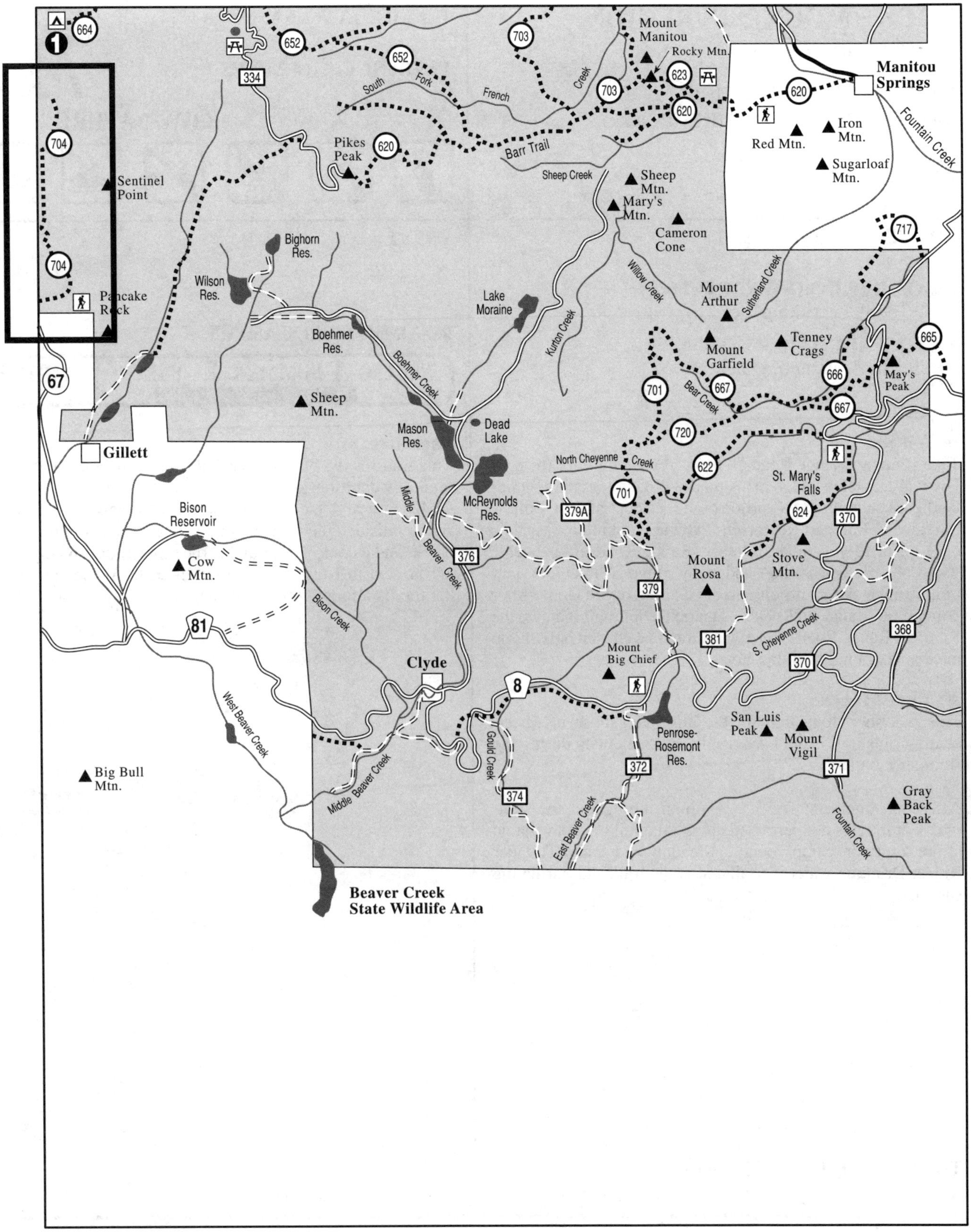

MAP 19

TRAIL #704

ROCKY MOUNTAIN REGION

MAP NUMBER(S): 16

NATIONAL FOREST: PIKE
RANGER DISTRICT: PIKES PEAK

USGS: CASCADE QUAD

DIFFICULTY: EASY

DISTANCE: 0.4 MILES

TRAIL 712

B.P.W. NATURE TRAIL

UPDATED: NOVEMBER 1991

USE: Light

TRAIL BEGINNING ELEVATION:
9,160 ft. at parking area

TRAIL ENDING ELEVATION:
9,180 ft. at parking area

RECOMMENDED SEASON:

SPRING	SUMMER	FALL	WINTER

ACCESS:
Take Rampart Range Road N. From Woodland Park (turn on Hwy 24 at McDonald's). This road will take you past Woodland Park schools and a portion of the residential area. Follow the signs to Rampart Reservoir. If you are looking for a scenic drive prior to your hike, Rampart Range Road offers scenic vistas of Colorado Springs and many canyons. Take the Manitou Springs exit as if going to the Garden of the Gods. After going past "Balanced Rock" a sign on the left will indicate the Rampart Range Road. This route should be chosen only if you have an extra hour driving time.

ATTRACTIONS:
This is a short loop trail for the blind and those in wheelchairs. This trail is only accessible to wheelchairs during the summer months.

Afternoon thunderstorms are common throughout the summer. Camping is not permitted along this trail but is available at the reservoir campgrounds: Meadow Ridge and Thunder Ridge. Drinking water is available at the trailhead during the summer.

narrative:
This trail for the physically challenged was built by the Colorado Federation of Business and Professional Women's Clubs Inc. and the Forest Service. It meanders through a sunny hillside, giant granite boulders surrounded by aspens, a creek, and a cool, wet fir and spruce forest. Information signs are printed in Braille, as well as English. At the entrance to the trail is a toilet with wheelchair access.

TRAIL PROFILE (ONE-WAY):

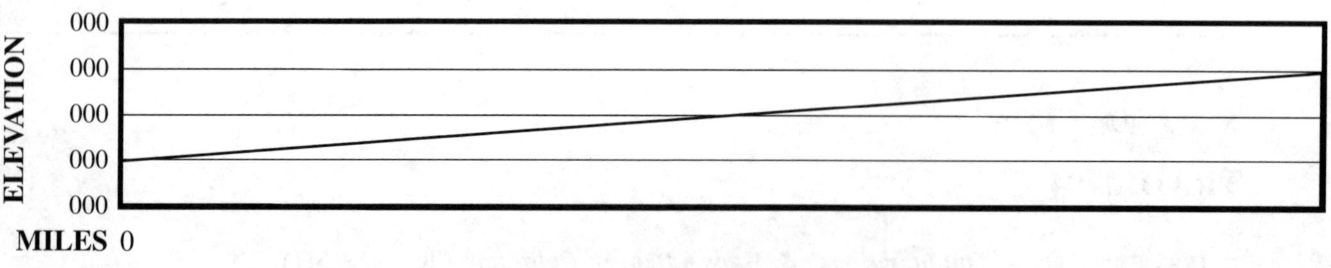

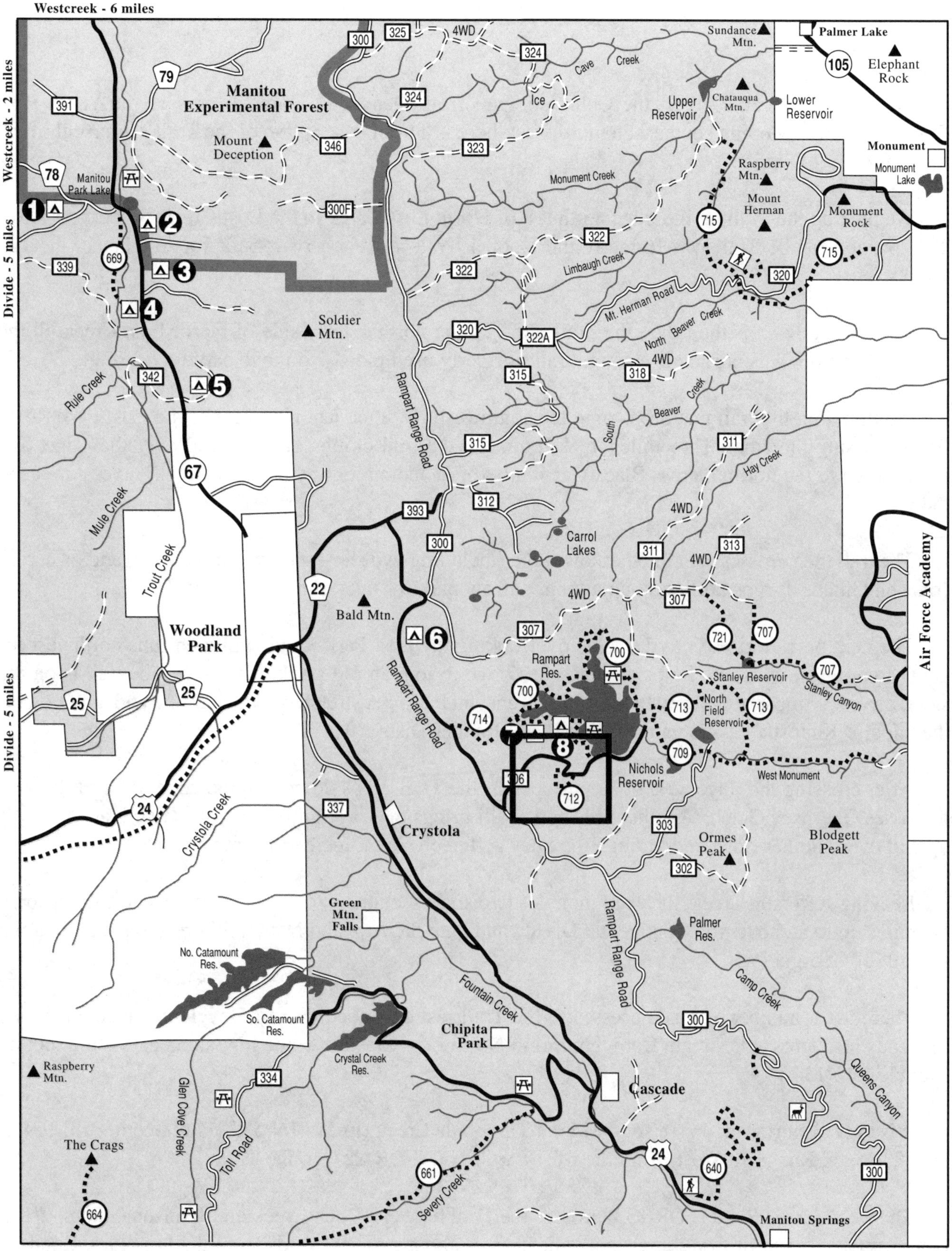

MAP 16

TRAIL #712

COLORADO TRAIL

For about 30 scenic miles, the Colorado Trail from Denver to Durango crosses the South Park Ranger District. Moving southeast to northwest, here is a brief description of the Trail's course through the South Park District.

We pick up the Trail on the ridge northeast of North Lost Creek in the Lost Creek Wilderness Area. This portion of the Trail was formerly the Hooper Trail, a wagon road built for logging in the late 1880's.

Dropping down off the ridge, the Trail turns up the northeasterly side of North Lost Creek all the way to its head. This portion is also old trail, formerly used mostly to move cattle.

We then join the trail that drops onto the southwesterly slope just northeast of the Lost Park Road. After crossing the small Threemile Creek drainage, the trail climbs back up and over the ridge, and drops down to the north to cross Black Canyon before winding its way to Rock Creek, several miles to the west.

The trail then crosses over another low ridge (including two short sections of old road) and on across Johnson Gulch. It runs northwest up a long, narrow park for over a mile.

The next portion rises onto the ridge overlooking the Lost Park Road; and then runs northwest for about two miles, where it climbs onto another ridge to join an old stock driveway. The trail is on the stock driveway most of the way to the top of Kenosha Pass. (A stretch of about a mile and a half on either side of Kenosha Pass is on the South Platte Ranger District.)

After crossing the ridge west of Kenosha Pass, the Trail drops down into the edge of South Park as it crosses Guernsey Gulch. Another mile and a half brings the traveler to Deadman Gulch; and then it's up and over another ridge and through the busy Jefferson Creek recreation area.

Leaving Jefferson Creek, the Trail includes two short sections of old road as it runs northwest some six miles up to and over the Continental Divide and onto the Arapaho National Forest to the road at the summit of Georgia Pass.

There are a number of road accesses to the Trail just described. Forest Development Road (FDR) 134 ends just across the stream from the trail in North Lost Creek. (Some low-slung cars might not be suitable for #134.)

It is only a short walk up the south side of Threemile Creek (in 5W 1/4 5W 1/4 of Section 13, Township 8 South, Range 73 West) from the Lost Park Road (FDR 127) to the Trail.

The Rock Creek Road (FDR 133) crosses the Trail between Rock Creek and Johnson Gulch. (#133 is not suitable for most low-slung cars.) The Trail is about two 200 yards from the Lost Park Road on top of the Dry Gulch Divide (NW 1,4 Section 12, Township 8 South, Range 75 West).

U.S. 285, of course, crosses the trail on top of Kenosha Pass. The Guernsey (FDR 809) & Deadman (FDR 427) Gulch Roads both reach the trail between Kenosha Pass and Jefferson Creek. (Neither is suited to low-slung passenger cars.)

The Jefferson Lake Road (FDR 401) crosses the Trail just north of Aspen Campground.

The Michigan Creek Road (FDR 400) meets the trail on top of Georgia Pass. (The upper 1/2-mile is not always suitable for low-slung passenger cars.)

The campgrounds fairly close to the Trail are Lost Park (2 miles south via the Brookside-McCurdy Trail 607), Kenosha Pass, and Aspen or Lodgepole in Jefferson Creek. Aside from these campgrounds, there is no developed drinking water along the Trail.

Dispersed camping is allowed throughout the area on this district that is traversed by the Trail except in Jefferson Creek, where camping and campfires are allowed only in campgrounds. Permits are not currently required for dispersed camping or campfires.

Several streams along the Colorado Trail offer fishing, including North Lost Creek, Black Canyon, Rock Creek and Jefferson Creek.

Deer and elk populate the country all along the Trail. Some bighorn sheep range in the Kenosha Mountains along North Lost Creek, as well as in the area easterly from Georgia Pass. There has been almost no problem of animals along the Trail route ever being a danger to man.

We do not encourage Colorado Trail travelers to bring their dogs along unless they are very carefully controlled. Dogs running loose can cause great stress to wildlife, horses, and sometimes other hikers.

Summer weather along the trail can be fickle, and travelers should be equipped for reasonably harsh conditions even in mid-summer. Frost (even occasional snow) can occur into mid-June and after mid-August. Heaviest thunderstorm activity usually runs from about July 10 through mid-August. Some high parts of the Trail near Georgia Pass are often still covered by snowdrifts in mid-June.

Noxious plants are virtually no problem in the Trail area. Only those who are very knowledgeable should gather and eat mushrooms, of course.

On this district, grades on the Colorado Trail exceed 8% only for short stretches. It is much less steep than many older trails. All of the Trail on this District is closed to motor vehicles.

In the general area of the Lost Creek Wilderness, the Colorado Trail connects with several others which can be incorporated into loop tours that last up to several days. The West Jefferson Trail will provide about an 11-mile loop tour in the Jefferson Creek area.

We wish all who visit the Colorado Trail a most enjoyable experience.

U.S. Forest Service

ROCKY MOUNTAIN REGION

MAP NUMBER(S): 4 and 5

NATIONAL FOREST: PIKE
RANGER DISTRICT: SOUTH PLATTE

USGS: KASSLER, PLATTE CANYON QUADS

DIFFICULTY: EASY

DISTANCE: 6.5 MILES

COLORADO TRAIL

1776A-B

KASSLER (WATERTON CANYON RECREATION AREA) TO STRONTIA SPRINGS DAM

UPDATED: AUGUST 1988

TRAIL BEGINNING ELEVATION:
5,600 ft.

TRAIL ENDING ELEVATION:
6,000 ft.

ACCESS:

Access to Kassler is by State Highway 75 near Kassler Water Treatment Plant (Waterton Canyon Recreation Area) Parking is available.

ATTRACTIONS:

Beginning in the parking lot at Waterton Canyon Recreation Area, visitors will find a wide variety of outdoor opportunities. Traveling by foot, horseback and bicycle invite mountain scenery close to the Denver Metro Area. This is a non-motorized trail section. Users should he advised that there is occasional vehicle travel on the Waterton Canyon Road due to operational and maintenance activities of the Denver Water Board. The canyon offers four-season recreation. Cross-country skiers, snowshoes and other winter visitors are welcome and encouraged. Photographers and those interested in nature study can view an environment with a variety of vegetation, wildlife and geologic formations. Species of plants and animals usually found in the higher elevations coexist with populations native to eastern Colorado Habitat is specifically managed to perpetuate the low altitude herd of Bighorn sheep. To protect the sheep, visitors must NOT bring dogs into the Waterton Canyon Recreation Area.

This portion of the South Platte River offers excellent fishing opportunities. From Strontia Springs Dam to 300 yards above the Marston Diversion has been designated Wild Trout Waters with fishing restrictions.

This narrow river canyon, through which the Denver, South Park and Pacific railroad ran, opened a shortcut to the rich South Park and Leadville mining camps. Many timber and tourist sites were located in the canyon. The existing service road runs along the old railroad grade.

Overnight camping and campfires are NOT permitted in Waterton Canyon

USE: Heavy.
HIKING, BIKING & HORSE TRAVEL ONLY.
NO MOTORIZED VEHICLES ALLOWED.

RECOMMENDED SEASON:

SPRING	SUMMER	FALL	WINTER

NARRATIVE:

None.

TRAIL PROFILE (ONE-WAY):

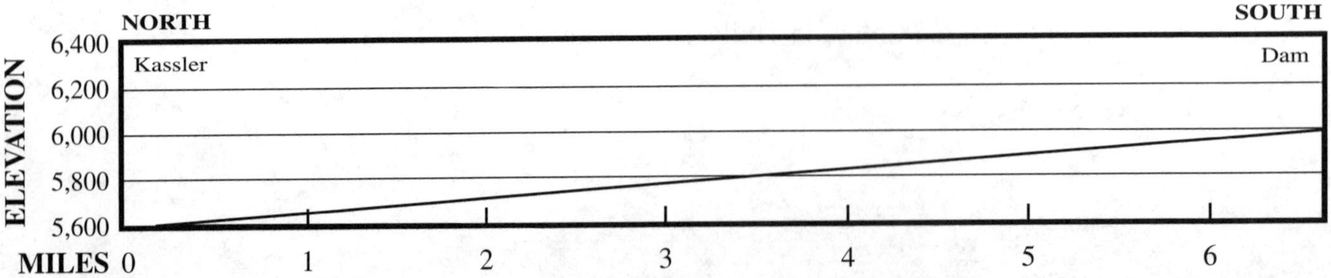

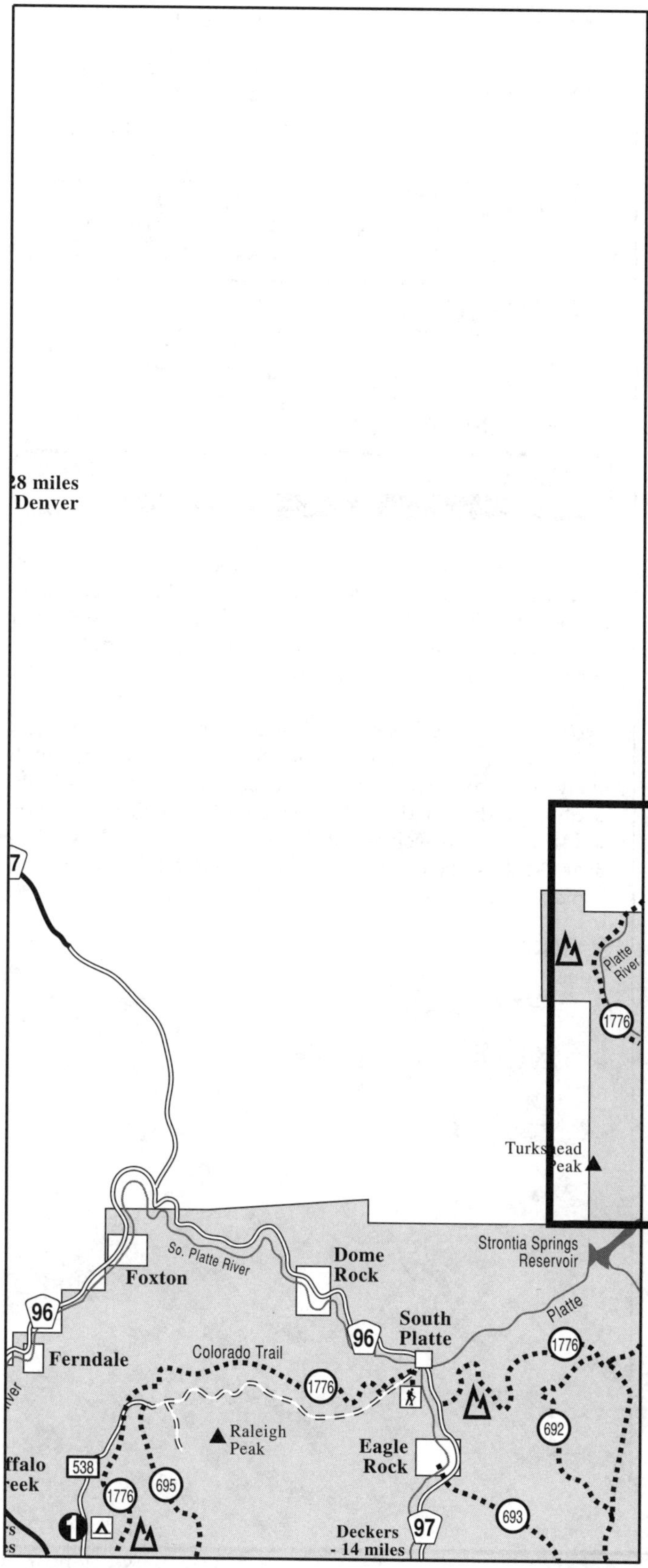

MAP 4 - East Half

TRAIL #1776 A-B

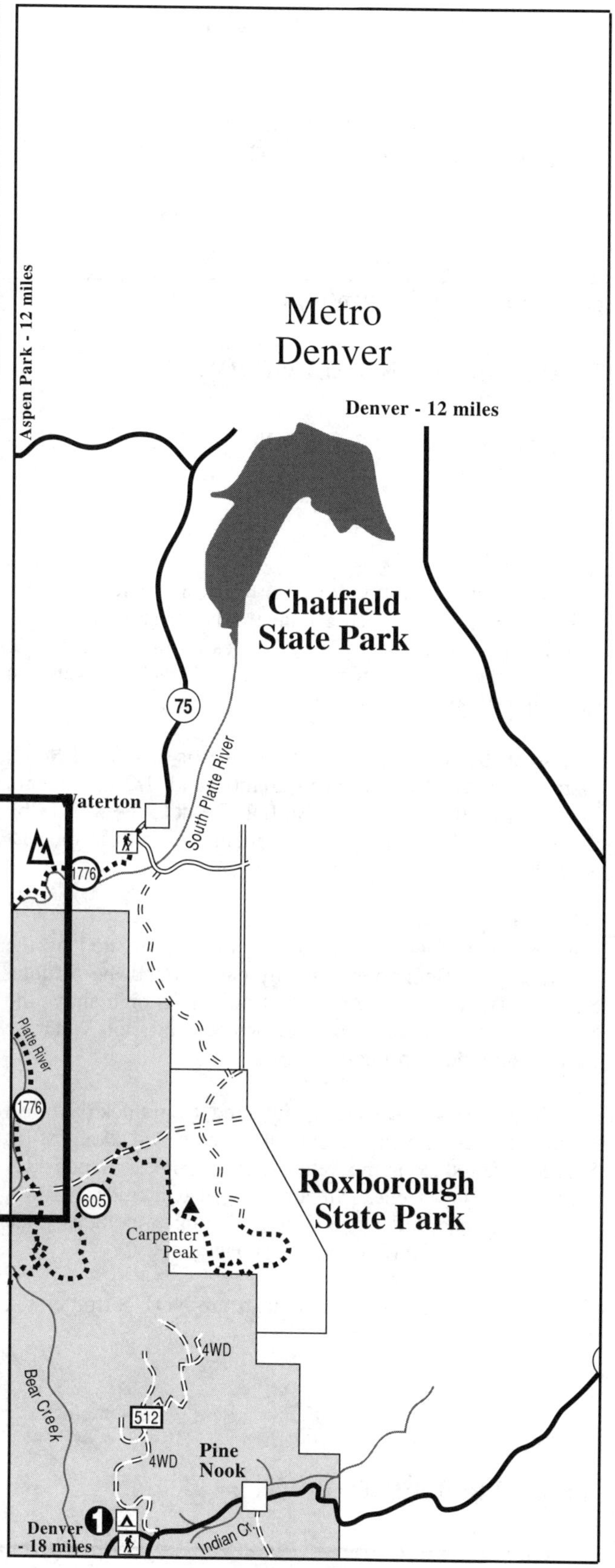

MAP 5 - West Half

ROCKY MOUNTAIN REGION

MAP NUMBER(S): 4 and 5

NATIONAL FOREST: PIKE
RANGER DISTRICT: SOUTH PLATTE

USGS: KASSLER, PLATTE CANYON QUADS

DIFFICULTY: MODERATE

DISTANCE: 10 MILES

COLORADO TRAIL 1776B-C

STRONTIA SPRINGS DAM TO OLD SOUTH PLATTE TOWNSITE

UPDATED: AUGUST 1988

TRAIL BEGINNING ELEVATION:
6,000 ft.

TRAIL ENDING ELEVATION:
6,380 ft.

USE: Light.
FOOT, BIKING & HORSE TRAVEL ONLY.
MOTORIZED VEHICLES PROHIBITED.

RECOMMENDED SEASON:

SPRING	SUMMER	FALL	WINTER

ACCESS:

This segment is from the Waterton Canyon Recreation Area trailhead at Kassler. Access to the trailhead is by State Highway 75. This segment begins 6.2 miles up the South Platte River at the Strontia Springs Dam. Travel in Waterton Canyon is 11 miles for foot, horse or bicycle.

Access at the south end of this segment is at the old South Platte townsite. The trail meets county road, 1/2 mile south. The townsite is reached from Buffalo Creek to the west or by the way of State Highway 67 and county roads from the east as shown on the vicinity map.

ATTRACTIONS:

This segment of the Colorado trail was constructed by the Denver Water Department as a by-pass around the Strontia Springs Dam and Reservoir. The trail climbs high above the South Platte river offering spectacular views of Waterton Canyon and the surrounding terrain.

A variety of ecosystems are encountered from thick oakbrush patches through stands of ponderosa pine and Douglas-fir. Spruce is found along the wetter stream courses. Aspen adds color to the fall scene. Streamside brush includes willow, mountain maple, alder and filbert. Wildlife includes mule deer, grouse, wild turkey and occasionally elk.

Except as posted, motorized vehicles are NOT permitted. A short section of the trail is used by motorized trail vehicles. Trail users, especially horseback riders, should be prepared to encounter motorcycles or ATV's. Barricades are installed at each end of that section to exclude motorized use from other portions of the trail . Tread Lightly.

The reservoir is accessible by way of an old undeveloped trail along Bear Creek. The reservoir shoreline is very steep with many sheer rock faces and is not recommended for fishing. Boating, wading, and swimming are prohibited. The dam area and structures, as well are off limits to the public .
Bear Creek is not suitable for fishing. The few fish which are found seldom are larger than 3 or 4 inches in length .

NARRATIVE:

None.

TRAIL PROFILE (ONE-WAY):

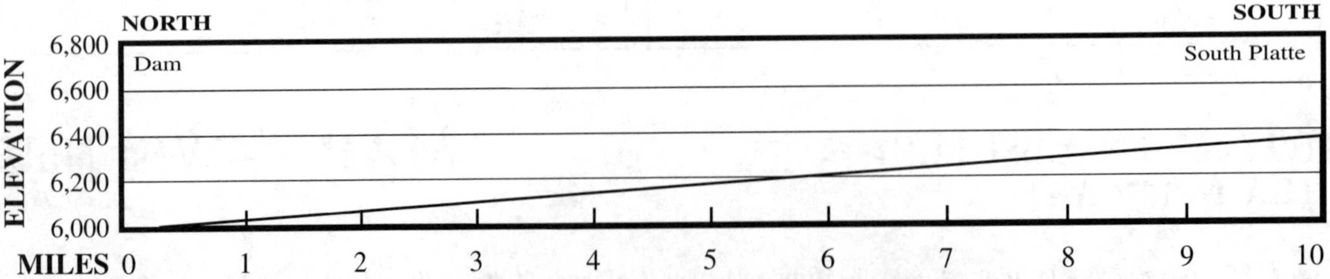

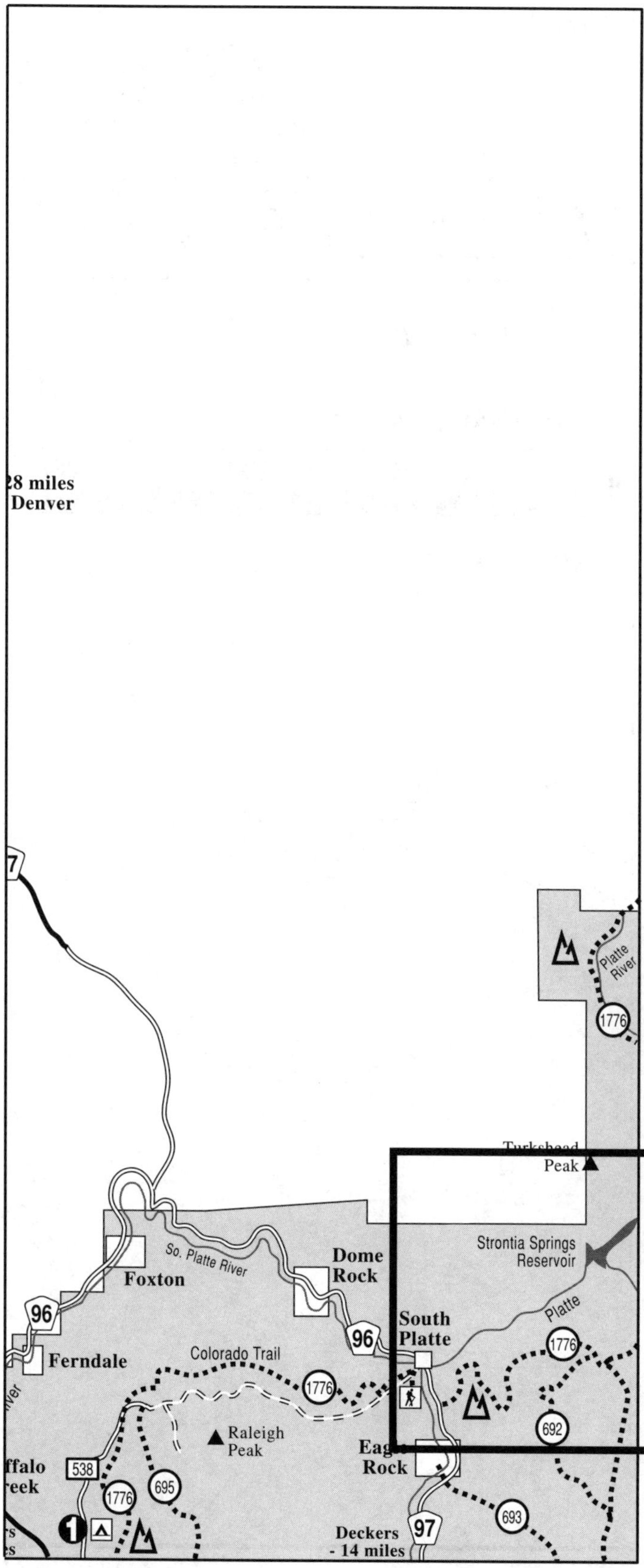

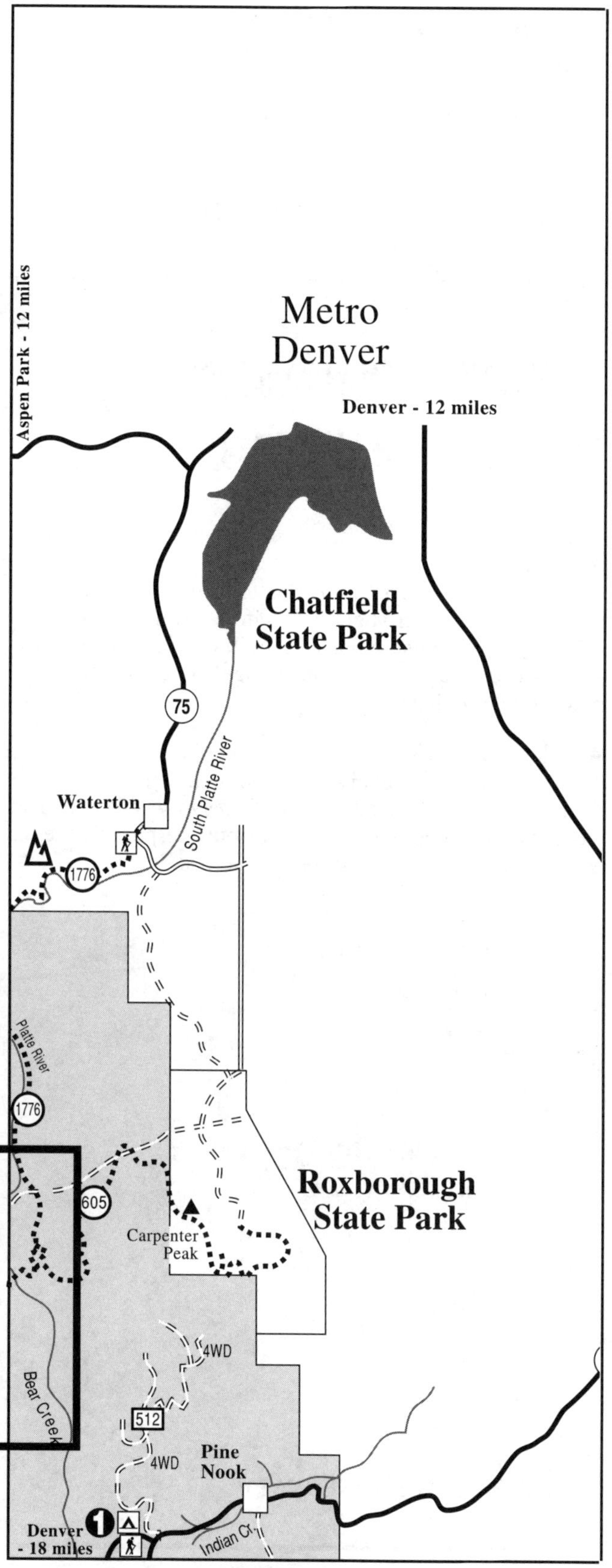

MAP 4 - East Half

TRAIL #1776 B-C

MAP 5 - West Half

ROCKY MOUNTAIN REGION
MAP NUMBER(S): 4

NATIONAL FOREST: PIKE
RANGER DISTRICT: SOUTH PLATTE

USGS: PLATTE CANYON QUAD

DIFFICULTY: EASY

DISTANCE: 0.8 MILES

COLORADO TRAIL 1776C-D

OLD SOUTH PLATTE TOWNSITE TO SOUTH PLATTE HOTEL

UPDATED: AUGUST 1988

TRAIL BEGINNING ELEVATION:
6,380 ft.

TRAIL ENDING ELEVATION:
6,380 ft.

ACCESS:
From U.S. highway 285, turn off onto Foxton Road, .06 miles southwest from traffic light in Conifer. Pavement ends after 4.7 miles. Continue another 3.4 miles to intersection with River Road. Turn east toward Longview and continue .05 miles to South Platte Hotel.

ATTRACTIONS:
This trail "segment' requires users to utilize the South Platte River Road in order to get across privately-owned land. This would be a good place for some "History" - especially the old South Platte Hotel and more on the railroad.

USE: Light.
HIKING & BIKING.

RECOMMENDED SEASON:

SPRING	SUMMER	FALL	WINTER

NARRATIVE:
None.

TRAIL PROFILE (ONE-WAY):

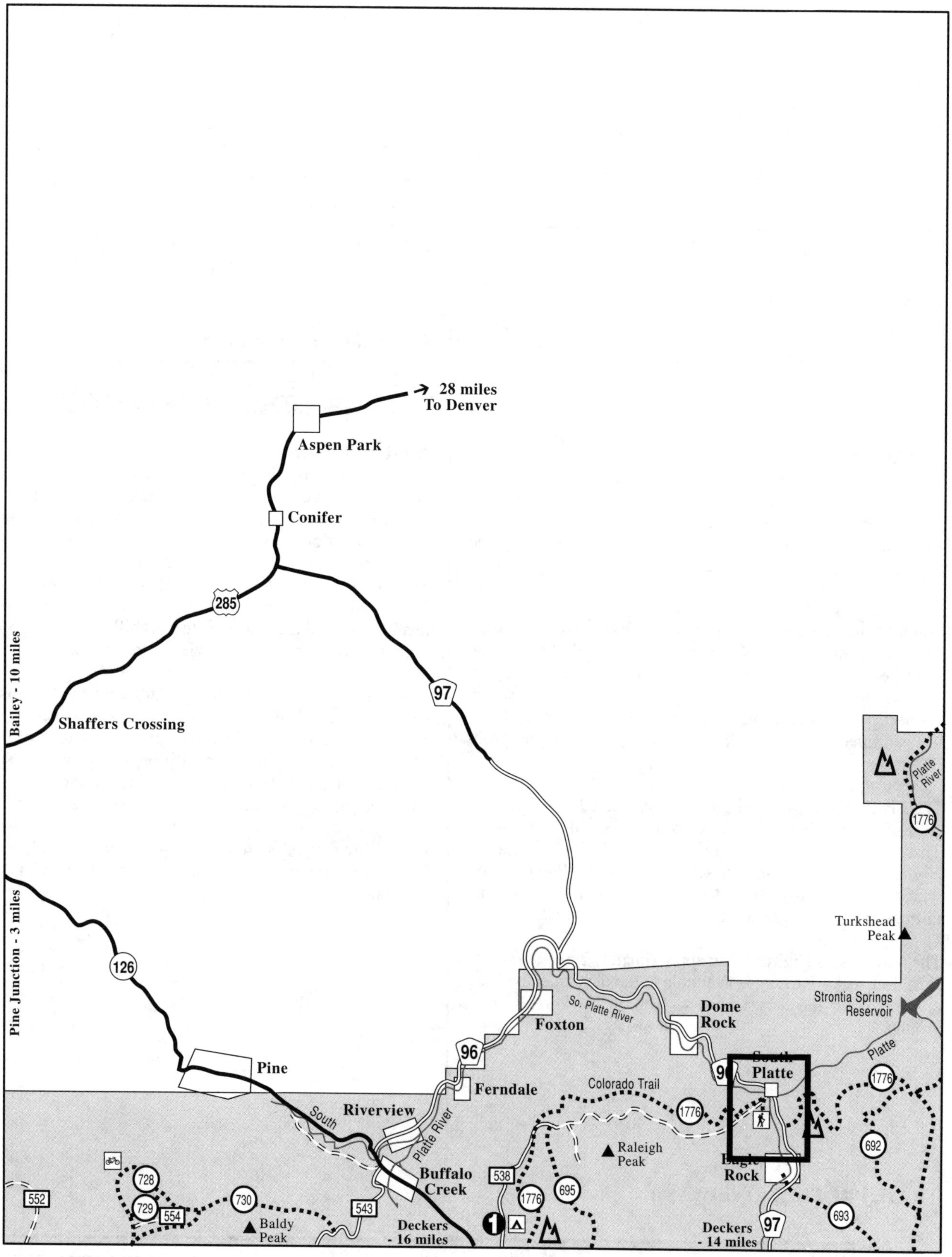

MAP 4
TRAIL #1776 C-D

ROCKY MOUNTAIN REGION

MAP NUMBER(S): 4 and 10

NATIONAL FOREST: PIKE
RANGER DISTRICT: SOUTH PLATTE

USGS: PLATTE CANYON, DECKERS QUADS

DIFFICULTY: MODERATE TO DIFFICULT
DISTANCE: 9.5 MILES

COLORADO TRAIL 1776D-E

SOUTH PLATTE HOTEL TO HIGHWAY 126 (BUFFALO CREEK)

UPDATED: AUGUST 1988

TRAIL BEGINNING ELEVATION:
6,380 ft.

TRAIL ENDING ELEVATION:
7,600 ft.

ACCESS:

Formal trailheads have not yet been constructed along this trail segment. Drinking water, toilets or overnight facilities are not provided. Parking areas are dependent on available space along roadsides and turnouts. Security is not provided for parked vehicles.

ATTRACTIONS:

South Platte town site is located 11 miles east of Buffalo Creek or 16 miles north of Deckers, on the South Platte River. Trail begins on the southwest side of the river following an old mining road. This is the beginning of Trail 695 which connects with Forest Road 538, the Rawlen Peak Road. Parking is limited at the site. Please respect posted private property.

The South Platte Railroad Station was located at the confluence of the North and South Forks of the South Platte River. There was a spur from here to the old mining camp of Nighthawk, and the station served the South Platte Hotel, which was generally full of tourists. This station was on the railroad line until 1938 when the line as dismantled.

The trail crosses Highway 126 approximately 3 miles south of Buffalo Creek. Parking is available at the developed parking area or along the access road to Top-of-the-World Campground.

USE: Light.
HIKING & BIKING.

RECOMMENDED SEASON:

SPRING	SUMMER	FALL	WINTER

NARRATIVE:

The terrain along this trail is steep and rocky with several peaks rising in close proximity to the trail. Many regular and intermittent streams are found throughout the area, however, this trail section does not cross any running water. The predominant type of vegetation is ponderosa pine with few grasses and forbes that survive in dry climates. Traveling west from South Platte, the trail goes uphill for two miles before leveling out. There is also a steep uphill climb to Top-of-the-World Campground, but the view is breathtaking.

This trail section passes through different land forms, from steep, gravel hillsides with little vegetation to gentle, rolling ridge-tops with outstanding scenery. If any user of the trail needs to be resupplied, whether it be food or water, supplies can be obtained at Buffalo Creek which is three miles north from the junction of the trail and Highway 126.

Forest Fire danger is often very high in this area. A good rule is to never smoke while traveling. Stop and to crush out your cigarette.

TRAIL PROFILE (ONE-WAY):

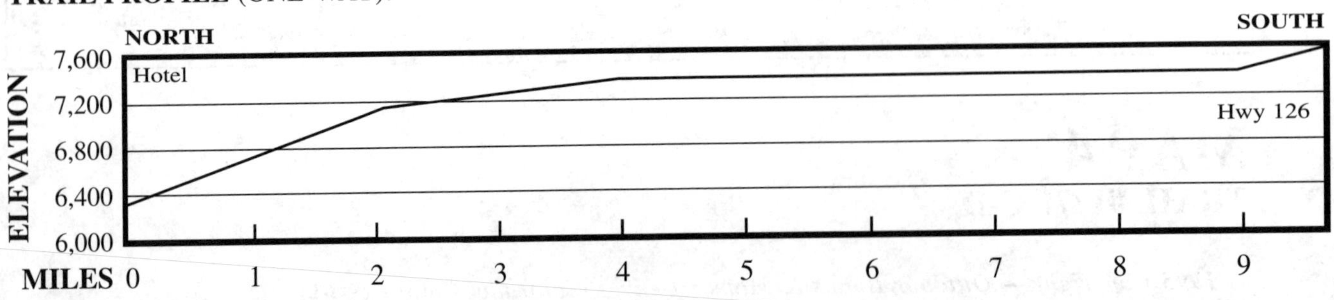

MAP 4 - South Half

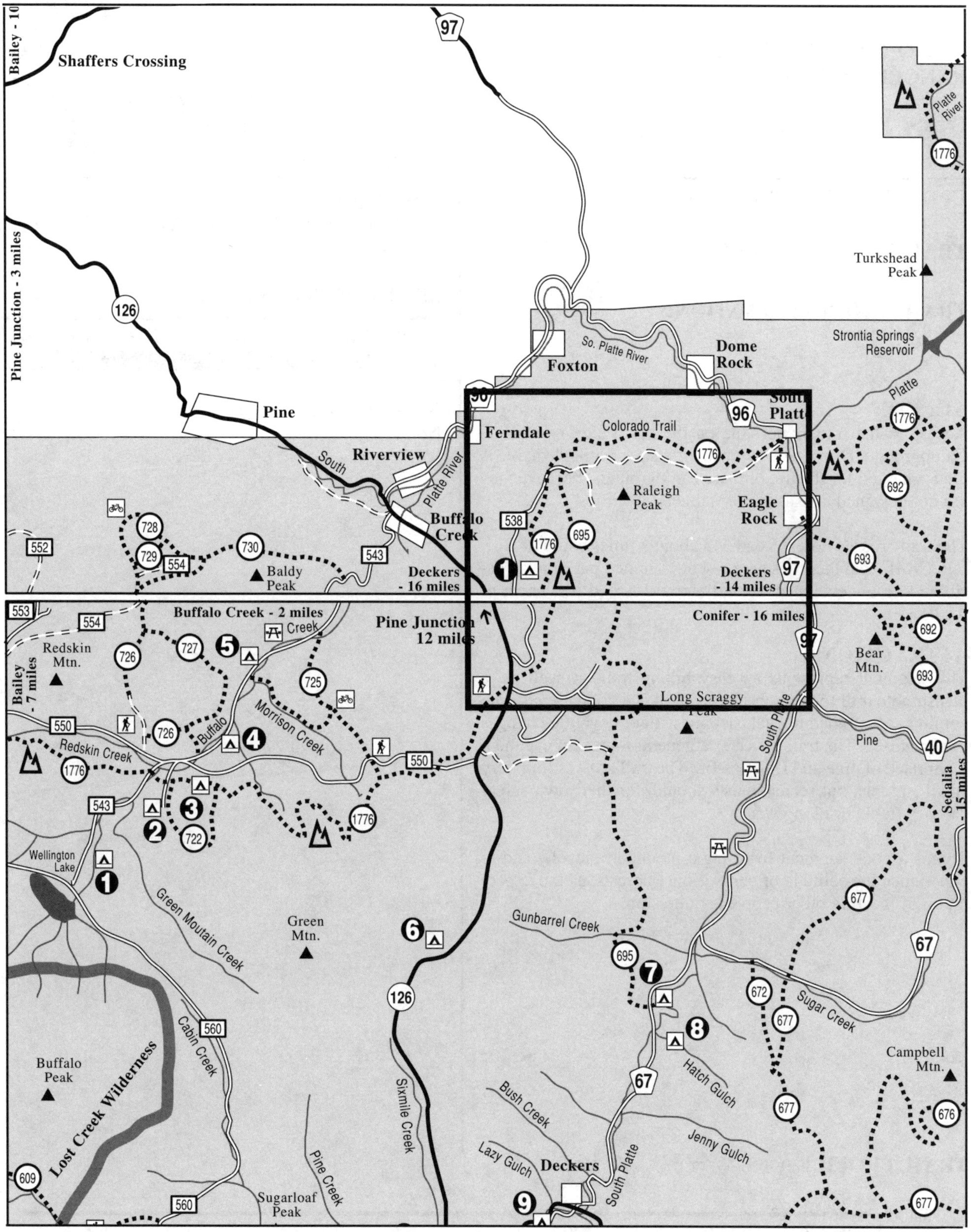

MAP 10 - North Half

TRAIL #1776 D-E

ROCKY MOUNTAIN REGION

MAP NUMBER(S): 10

NATIONAL FOREST: PIKE
RANGER DISTRICT: SOUTH PLATTE

USGS: DECKERS, GREEN MOUNTAIN QUADS

DIFFICULTY: EASY

DISTANCE: 9 MILES

COLORADO TRAIL 1776E-F

HIGHWAY 126 TO THE CROSSING OF WELLINGTON LAKE ROAD #543

UPDATED: AUGUST 1988

TRAIL BEGINNING ELEVATION:
7,600 ft.

TRAIL ENDING ELEVATION:
8,350 ft.

ACCESS:

Access to the trail is possible on the south side of Buffalo Campground. Drinking water and toilets are available here. A fee is charged for the use of these campgrounds, because they are not designed as parking or trailhead facilities.

The trail crosses Forest Road 543 about 7 miles south of Buffalo Creek. No facilities are present. Limited parking is possible at an undeveloped turnout along the road just north of the crossing

ATTRACTIONS:

This segment represents an easy hiking trail, as well as, a pleasurable trail for horseback travel. It is a good area for becoming accustomed to backpacking with only gentle changes in elevation . The trail traverses a general forest environment comprised of pine and Douglas-fir. The trail crosses four very small streams, but recreationists should, for their own safety carry portable drinking water.

This is a good segment for using a mountain bicycle, and to experience the solitude of our National Forests, as the expectation of meeting other people is quite slim.

USE: Light.
HIKING, BIKING & HORSEBACK RIDING.

RECOMMENDED SEASON:

SPRING	SUMMER	FALL	WINTER

NARRATIVE:

None.

TRAIL PROFILE (ONE-WAY):

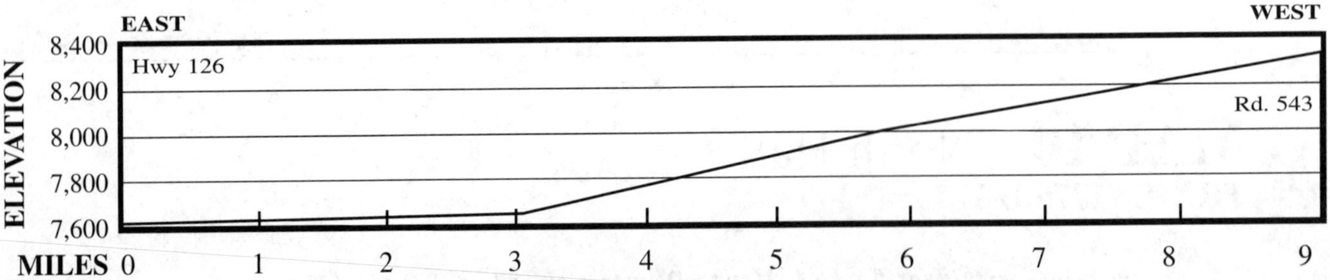

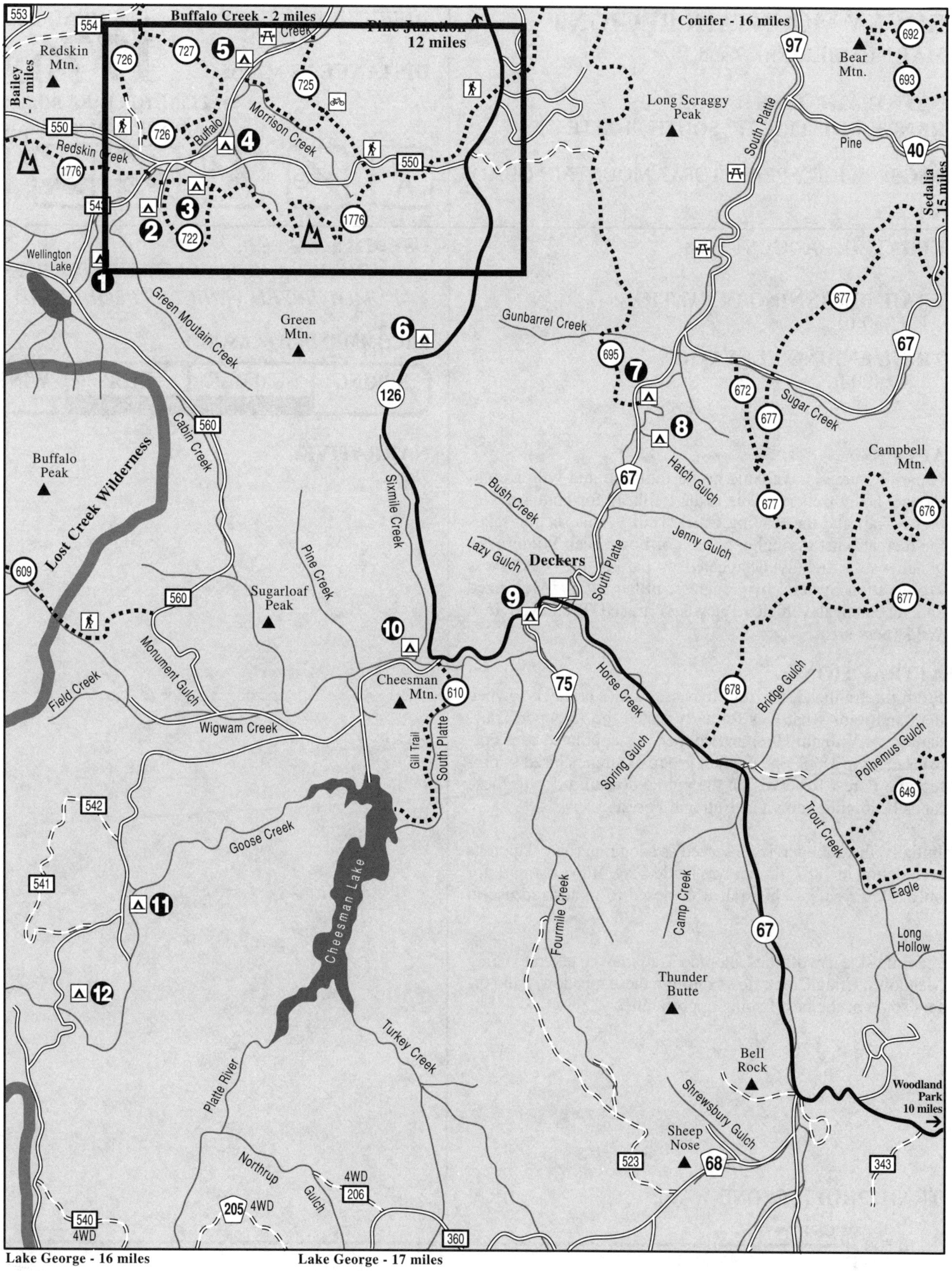

MAP 10

TRAIL #1776 E-F

ROCKY MOUNTAIN REGION

MAP NUMBER(S): 9 and 10

NATIONAL FOREST: PIKE
RANGER DISTRICT: SOUTH PLATTE

USGS: WINDY PEAK, TOPAZ MOUNTAIN QUADS

DIFFICULTY: EASY

DISTANCE: 8 MILES

COLORADO TRAIL
1776F-G
WELLINGTON LAKE ROAD #543 TO LOST PARK CAMPGROUND TRAIL

UPDATED: AUGUST 1988

TRAIL BEGINNING ELEVATION:
8,350 ft.

TRAIL ENDING ELEVATION:
10,400 ft.

ACCESS:
A parking area is available at the trailhead and will accommodate horse trailers. This is the trailhead for both the Colorado Trail and the Rolling Creek Trail . This section takes the recreationist through a portion of Lost Creek Wilderness, which was officially designated as part of the Wilderness Preservation System, by Congress, in late 1980. Motorized and mechanical vehicles (bicycles) are NOT allowed in a Wilderness area.

ATTRACTIONS:
From the trailhead, the trail bypasses some privately-owned land and joins what was formerly called the "Hooper Trail" named for William Hooper. Hooper was appointed as a Forest Ranger in 1898, one the first group of rangers hired to protect the Forest Reserves on the public domain prior to these lands becoming known as National Forests.

Initially, the 'Hooper Trail' served as a logging road. Upon its completion in 1887, timber was hauled over it to Lost Park for only a few years. This trail was closed to motorized use in 1974.

A 2 mile side trip off the Colorado Trail may be taken to Craig Meadows. Craig Creek flows through these meadows, and offers some of the best fishing opportunities.

USE: Light.
HIKING & HORSEBACK RIDING. BIKES OR MOTORIZED VEHICLES PROHIBITED

RECOMMENDED SEASON:

SPRING	SUMMER	FALL	WINTER

NARRATIVE:
None.

TRAIL PROFILE (ONE-WAY):

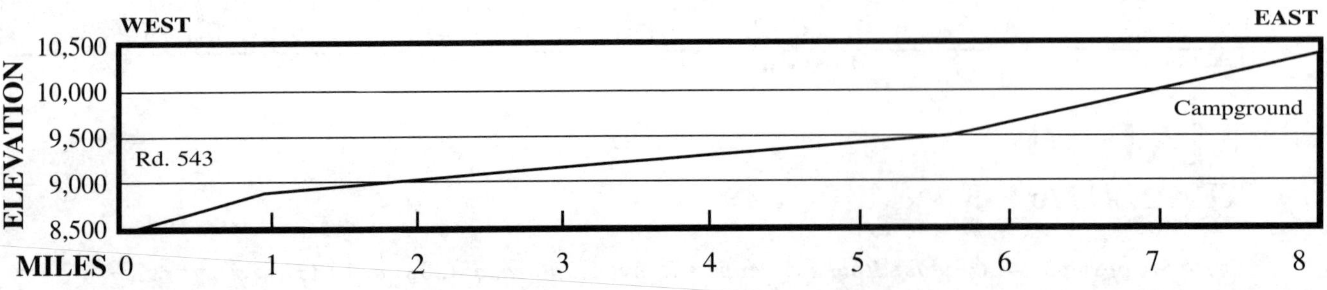

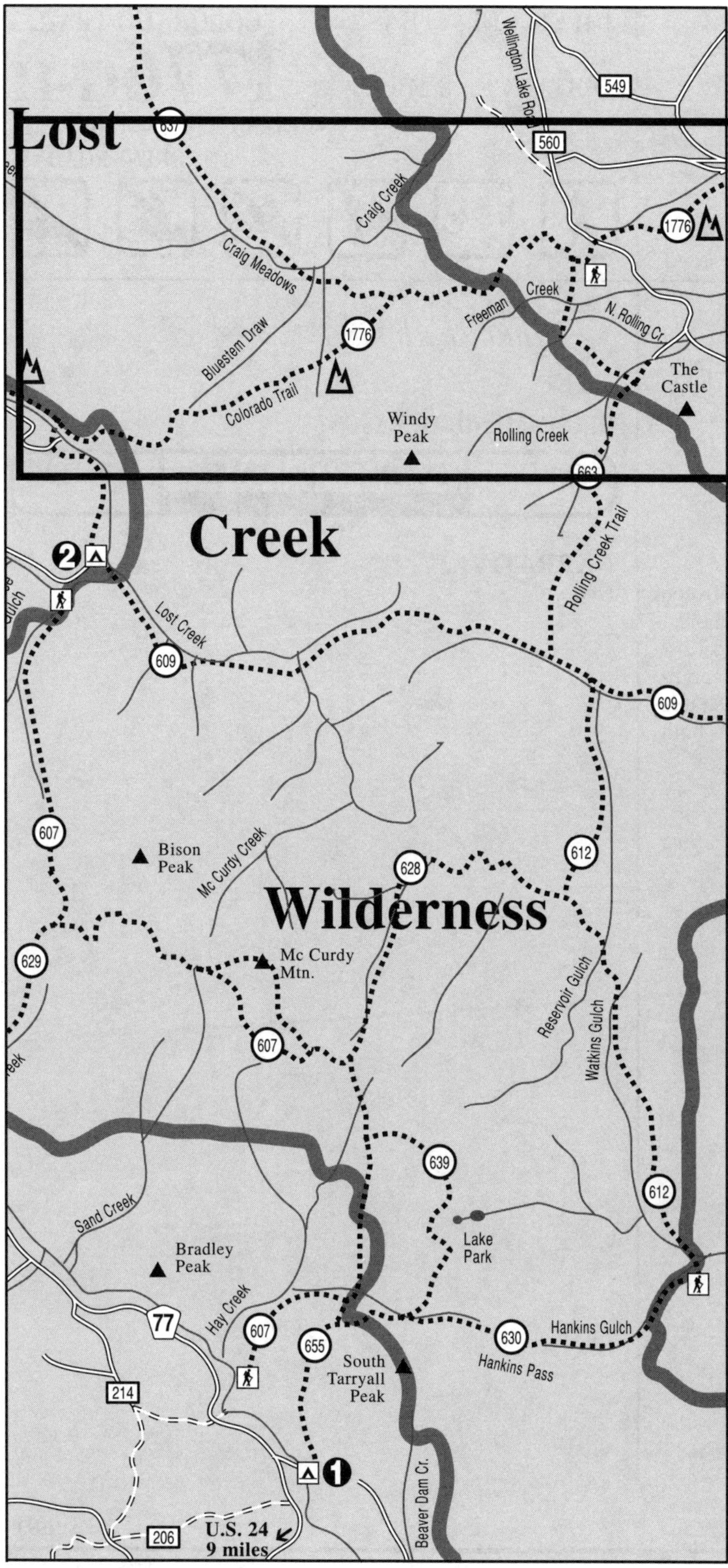

MAP 9 - East Half

TRAIL #1776 F-G

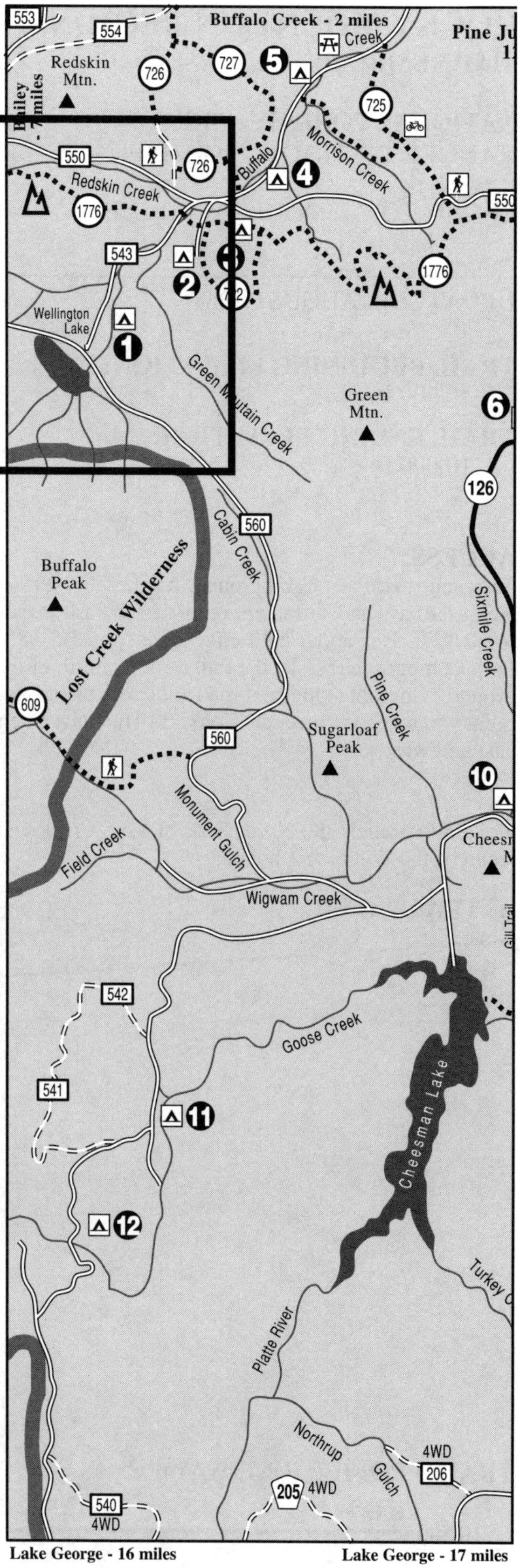

MAP 10 - West Half

ROCKY MOUNTAIN REGION
MAP NUMBER(S): 9

NATIONAL FOREST: PIKE
RANGER DISTRICT: SOUTH PLATTE

USGS: TOPAZ MOUNTAIN QUAD

DIFFICULTY: EASY

DISTANCE: 8 MILES

COLORADO TRAIL 1776G-H

LOST PARK CAMPGROUND TRAIL TO FOREST ROAD #56

UPDATED: AUGUST 1988

TRAIL BEGINNING ELEVATION:
10,400 ft.

TRAIL ENDING ELEVATION:
10,000 ft.

ACCESS:
To reach Lost Park Campground, turn off US 285. between Kenosha Pass and Jefferson, on the Lost Park Roads Forest Road 127. It is about a 20 mile drive from US 285 to Lost Park Campground. Take the trail directly north of the campground. Go right at the next road and cross the creek. At this point you will soon meet the Colorado Trail as it extends both east and west.

The trail parallels the North Fork of Lost Creek and is just south of the Wilderness boundary.

ATTRACTIONS:
None.

USE: Light.
HIKING & BIKING.

RECOMMENDED SEASON:

SPRING	SUMMER	FALL	WINTER
	■	■	

NARRATIVE:
None.

TRAIL PROFILE (ONE-WAY):

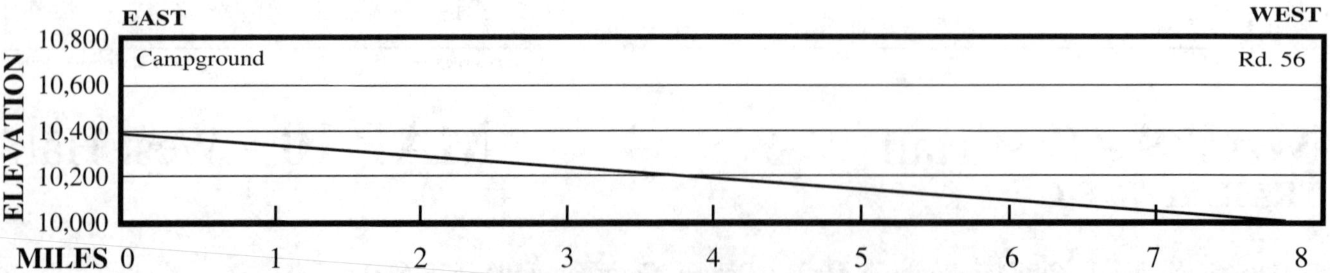

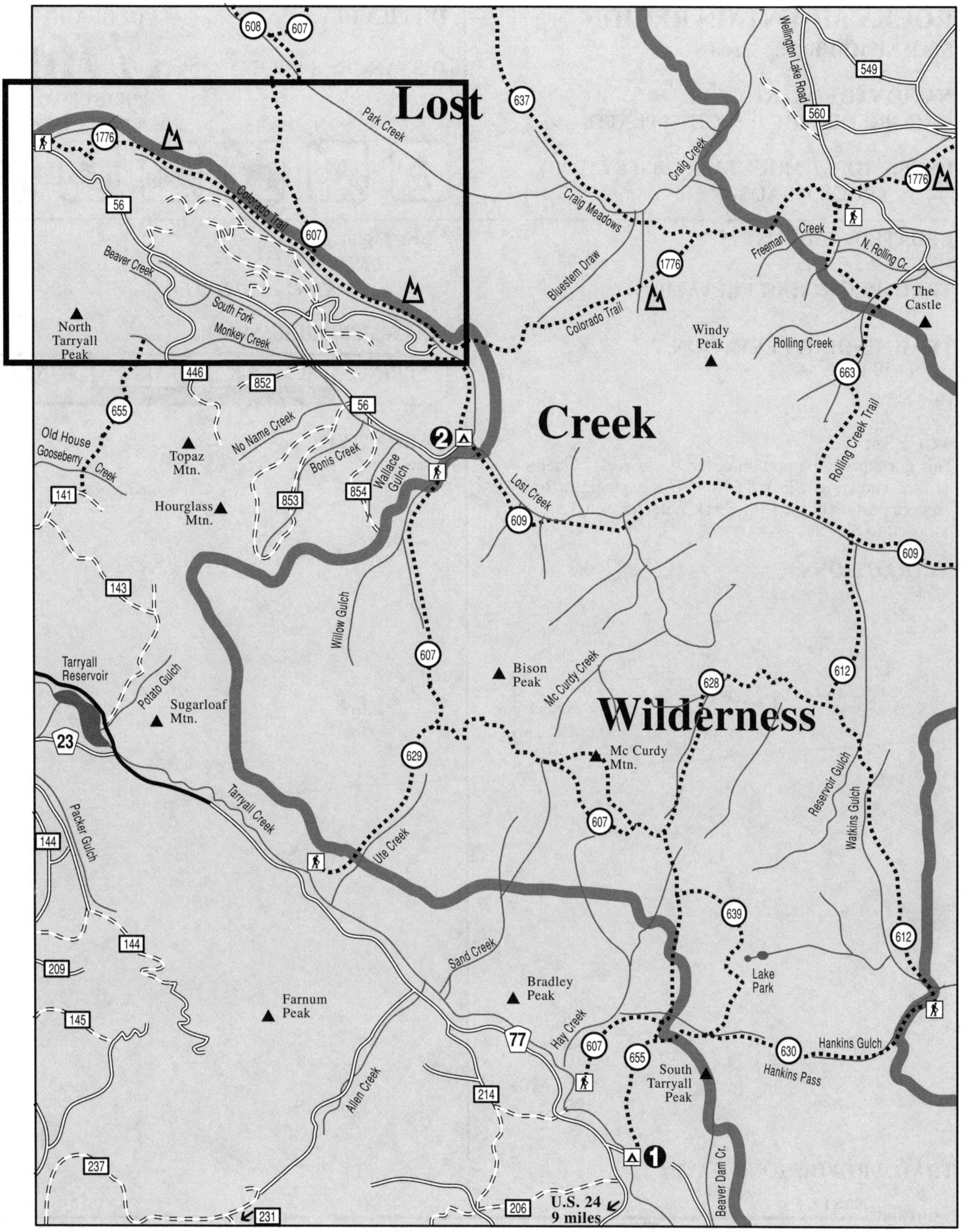

MAP 9

TRAIL #1776 G-H

ROCKY MOUNTAIN REGION

MAP NUMBER(S): 2 and 8

NATIONAL FOREST: PIKE
RANGER DISTRICT: SOUTH PLATTE

USGS: TOPAZ MOUNTAIN, OBSERVATORY ROCK QUADS

DIFFICULTY: EASY

DISTANCE: 8 MILES

COLORADO TRAIL
1776H-I
FOREST ROAD #56 TO ROCK CREEK COW CAMP

UPDATED: AUGUST 1988

TRAIL BEGINNING ELEVATION:
10,000 ft.

TRAIL ENDING ELEVATION:
9,650 ft.

ACCESS:
This goes through a portion of the Lost Creek Wilderness in an area known as Black Canyon. This segment of trail traverses a portion of the Lost Creek Wilderness and is closed to bicycles

ATTRACTIONS:
None.

USE: Light.
HIKING ONLY.
NO BIKING PERMITTED.

RECOMMENDED SEASON:

SPRING	SUMMER	FALL	WINTER
	■	■	

NARRATIVE:
None.

TRAIL PROFILE (ONE-WAY):

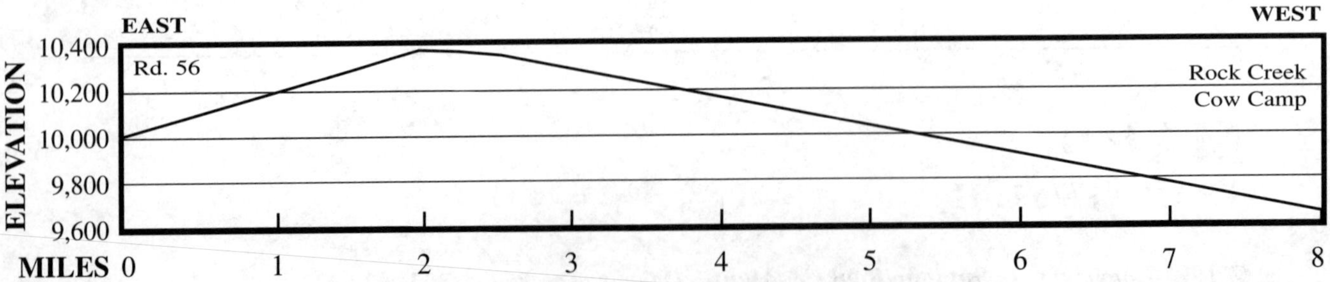

MAP 2 - South Half

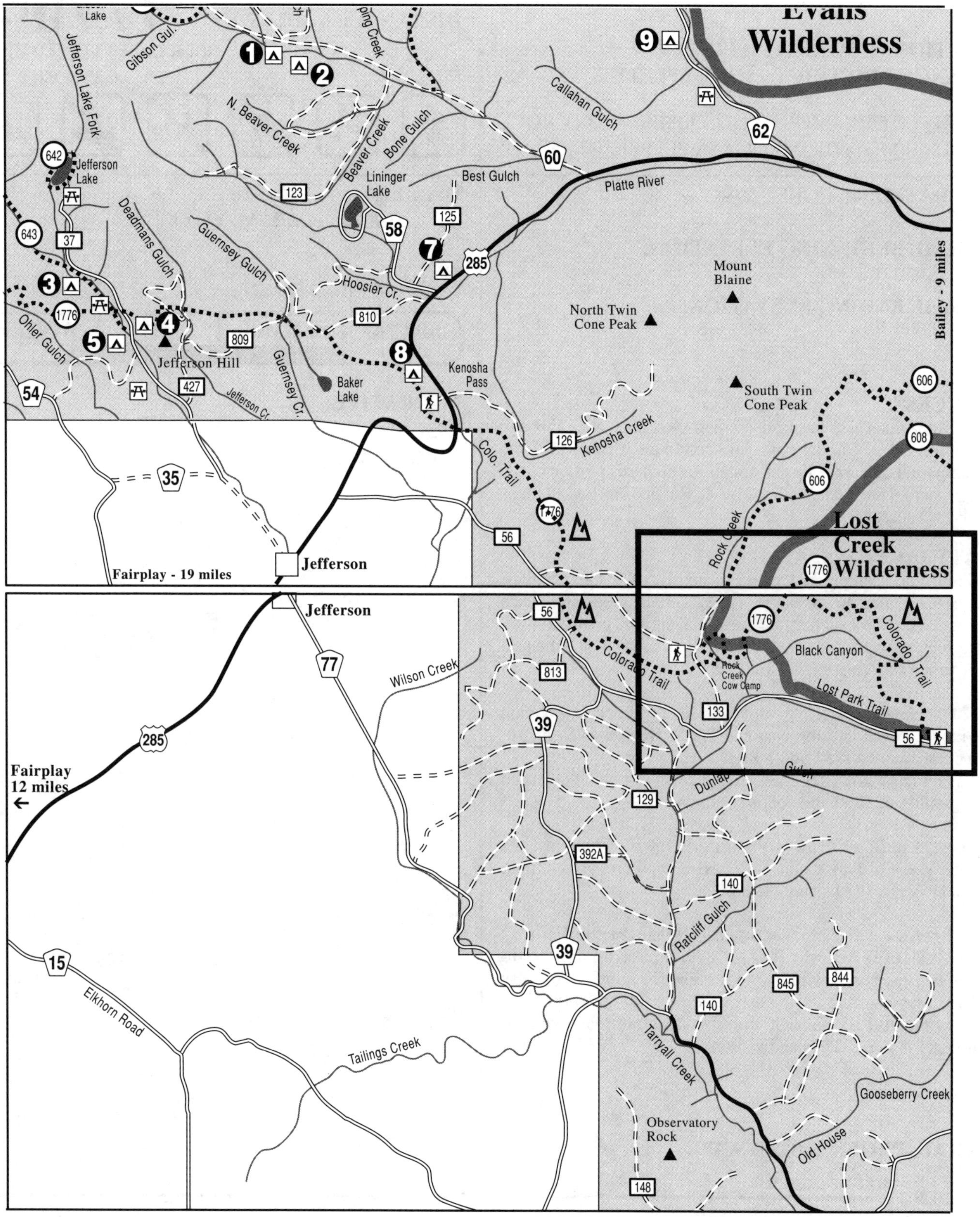

MAP 8 - North Half

TRAIL #1776 H-I

ROCKY MOUNTAIN REGION

MAP NUMBER(S): 2 and 8

NATIONAL FOREST: PIKE
RANGER DISTRICT: SOUTH PLATTE

USGS: TOPAZ MOUNTAIN, OBSERVATORY ROCK, MOUNTAIN LOGAN& JEFFERSON QUADS

DIFFICULTY: EASY

DISTANCE: 6 MILES

COLORADO TRAIL

1776I-J

ROCK CREEK COW CAMP TO KENOSHA PASS

UPDATED: AUGUST 1988

TRAIL BEGINNING ELEVATION:
9,650 ft.

TRAIL ENDING ELEVATION:
9,994 ft.

USE: Light.
HIKING & BIKING ONLY.

RECOMMENDED SEASON:

SPRING	SUMMER	FALL	WINTER

ACCESS:

At Kenosha Pass, the trail crosses U.S. 285, goes through Kenosha Pass Campground and continues westward to the Jefferson Lake area. For information about this portion of the Colorado Trail contact the District Ranger's office in Fairplay, Colorado.

ATTRACTIONS:

South Park comprises an area of nearly 1000 square miles. One can imagine the large varied animal herds that once roamed the Park such as the buffalo. This area provided a superb hunting ground for the Ute Indians as well as their prehistoric predecessors.

The abundance of beaver attracted the fur trappers. Zebulon Pike (the peak and the National Forest are named in his remembrance) crossed South Park in 1806.
The miners came after 1859. A great deal of mining occurred beginning in 1860 and continues today.

Located in South Park is the historic town of Fairplay, the county seat of Park County. It was an important gold mining center in the 1870's and 1880's.

The narrow gauge Denver South Park and Pacific Railroad was built over Kenosha Pass in 1879. In South Park, ranching was undertaken within a year of the first gold strikes. By 1883, there were early 50,000 head of cattle grazing in the Park. South Park has held the fascination of many generations of both residents and visitors.

NARRATIVE:

None.

TRAIL PROFILE (ONE-WAY):

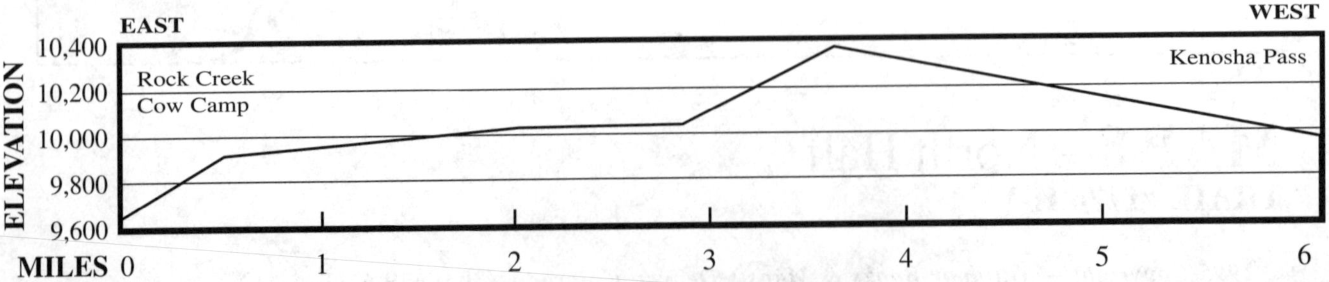

MAP 2 - South Half

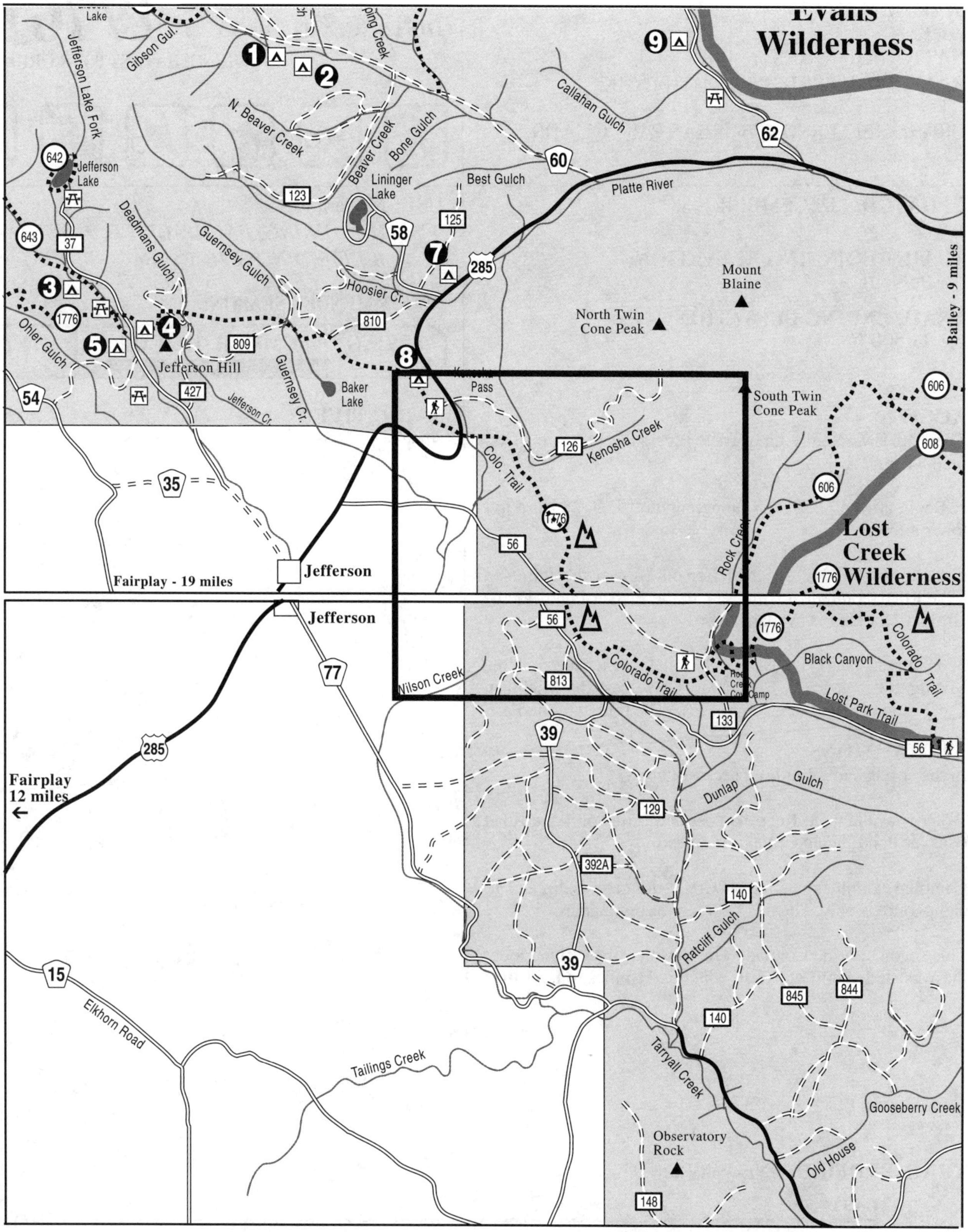

MAP 8 - North Half

TRAIL #1776 I-J

ROCKY MOUNTAIN REGION

MAP NUMBER(S): 1 and 2

NATIONAL FOREST: PIKE
RANGER DISTRICT: SOUTH PARK

USGS: JEFFERSON, BOREAS PASS QUADS

DIFFICULTY: MODERATE

DISTANCE: 11 MILES

COLORADO TRAIL 1776J-K

KENOSHA PASS TO GEORGIA PASS

UPDATED: DECEMBER 1988

TRAIL BEGINNING ELEVATION:
9,994 ft.

TRAIL ENDING ELEVATION:
11,500 ft.

USE: Moderate.
DAY HIKING, BACKPACKING, MOUNTAIN BIKE RIDING & HORSEBACK RIDING.

RECOMMENDED SEASON:

SPRING	SUMMER	FALL	WINTER
	■	■	

ACCESS:
There are several different ways to access this part of the Colorado Trail:

1. Begin at Kenosha Pass Campground off U.S. 285 and follow trail west.

2. Continue on U.S. 285 to Jefferson Turn at the Jefferson Lake Road, continue 5.2 miles to the Beaver Ponds Picnic Ground. Parking is available here.

3 . Start at Georgia Pass which can be reached from the Michigan Creek Road, FDR 400. Drive past the Michigan Creek Campground to Georgia Pass The trail is to the East.

ATTRACTIONS:
In the spring snow can limit accessibility.

Water is available at Kenosha Pass and Jefferson Lake Road (Beaver Pond Picnic Ground) trailheads.

Camping within the Jefferson Creek drainage is limited to campgrounds only. There is a $6 fee for these areas.

This segment of he Colorado Trail has the first official loop, the West Jefferson Loop Trail which is 11 miles round trip.

NARRATIVE:
None.

TRAIL PROFILE (ONE-WAY):

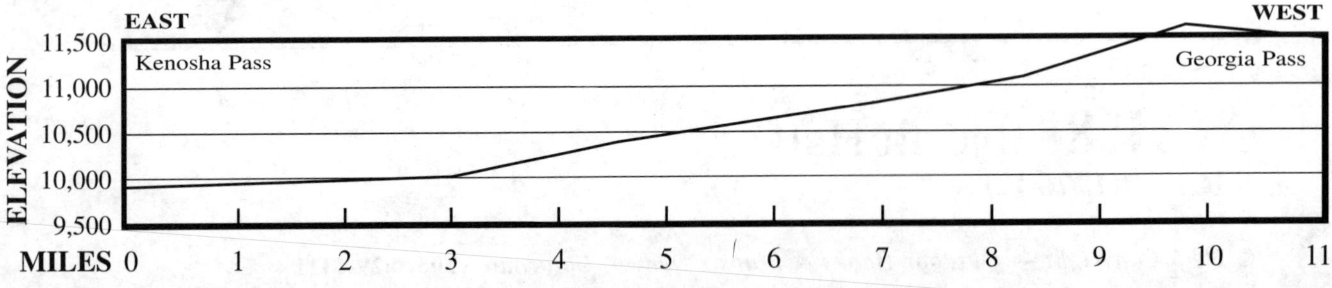

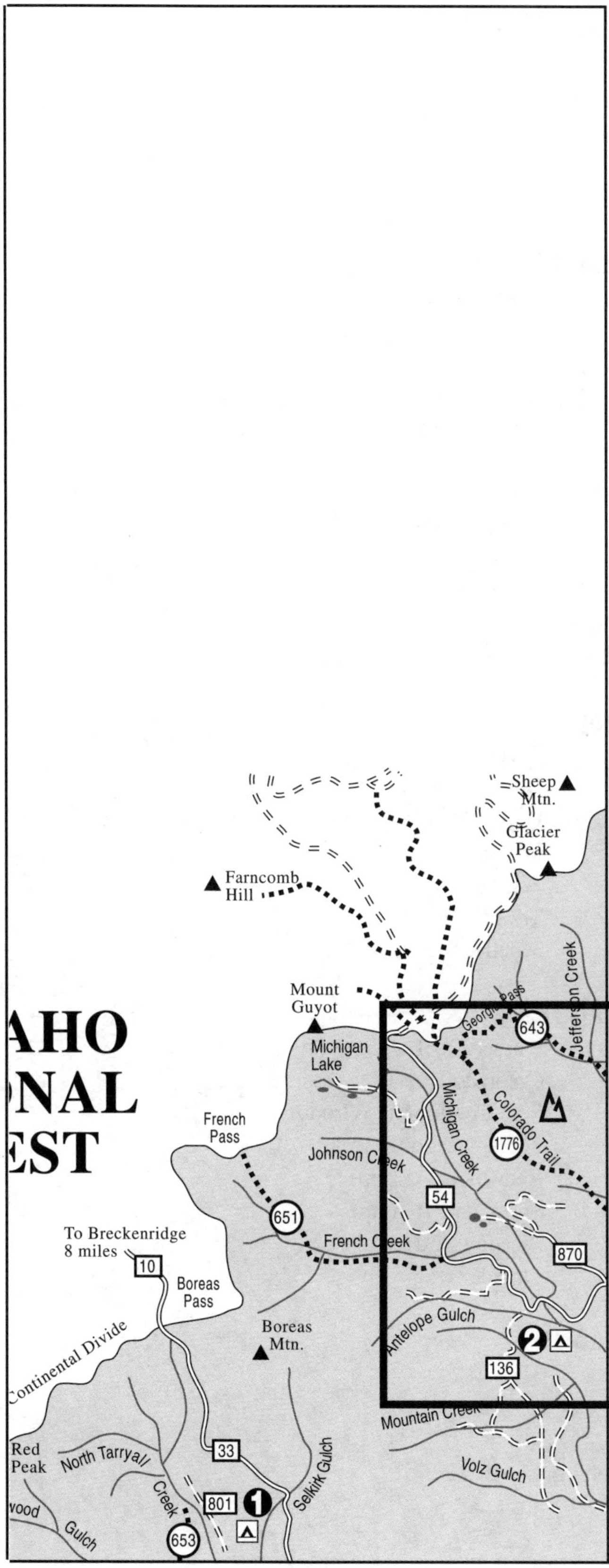

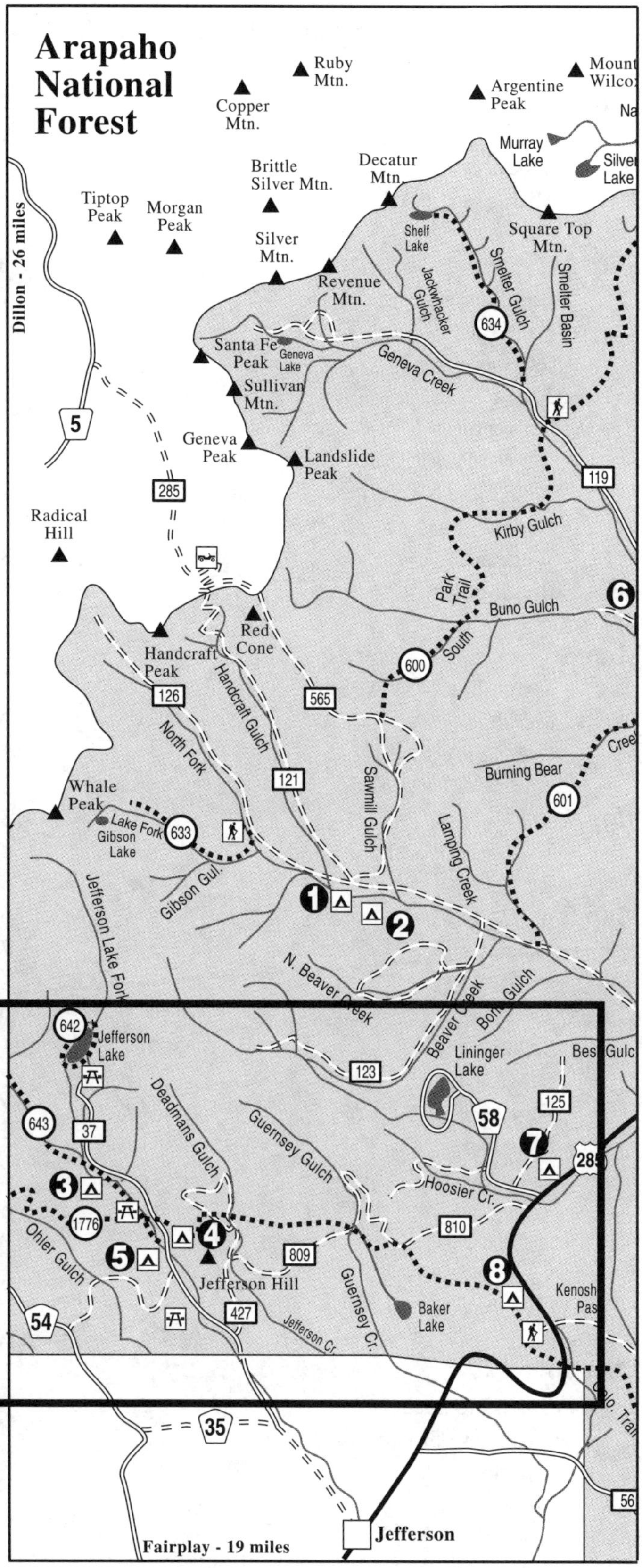

MAP 1 - East Half

TRAIL #1776 J-K

MAP 2 - West Half

Campgrounds Pike National Forest

Map 1

Site #	Name
1.	Selkirk
2.	Michigan Creek

Map 2

1. Hall Valley
2. Hand Cart
3. Jefferson Creek
4. Aspen
5. Lodgepole
6. Geneva Park
7. Timberline (Group)
8. Kenosha Pass
9. Whiteside
10. Burning Bear

Map 3

1. Mount Evans SWA*
2. Meridian
3. Deer Creek

Map 4

1. Top of the World

Map 5

1. Indian Creek

Map 7

1. Selkirk
2. Beaver Creek
3. Four Mile
4. Horseshoe
5. Kite Lake

Map 9

1. Spruce Grove
2. Lost Park

Map 10

1. Green Mountain
2. Meadows (Group)
3. Buffalo
4. Tramway
5. Baldy
6. Kelsey
7. Platte River
8. Bridge Crossing
9. Lone Rock
10. Wigwam
11. Molly Gulch
12. Goose Creek

Map 11

1. Flat Rocks
2. Devils head
3. Jackson Creek

Map 12

1. Weston Pass

Map 13

1. Buffalo Spring

Map 14

1. Round Mountain
2. Springer Gulch
3. Wagon tongue
4. Cove
5. Spillway

7-13 Elevenmile State Park*

Map 15

1. Riverside
2. Blue Mountain
3. Happy Meadows
4. Wildhorn
5. Big Turkey
6. Trail Creek
7. Mueller State Park*

Map 16

1. Painted Rocks
2. Colorado
3. Pike Community (Group)
4. South Meadows
5. Red Rocks (Group)
6. Springdale (Group)
7. Meadow Ridge
8. Thunder Ridge

Map 19

1. The Crags

* Will not appear on camping chart

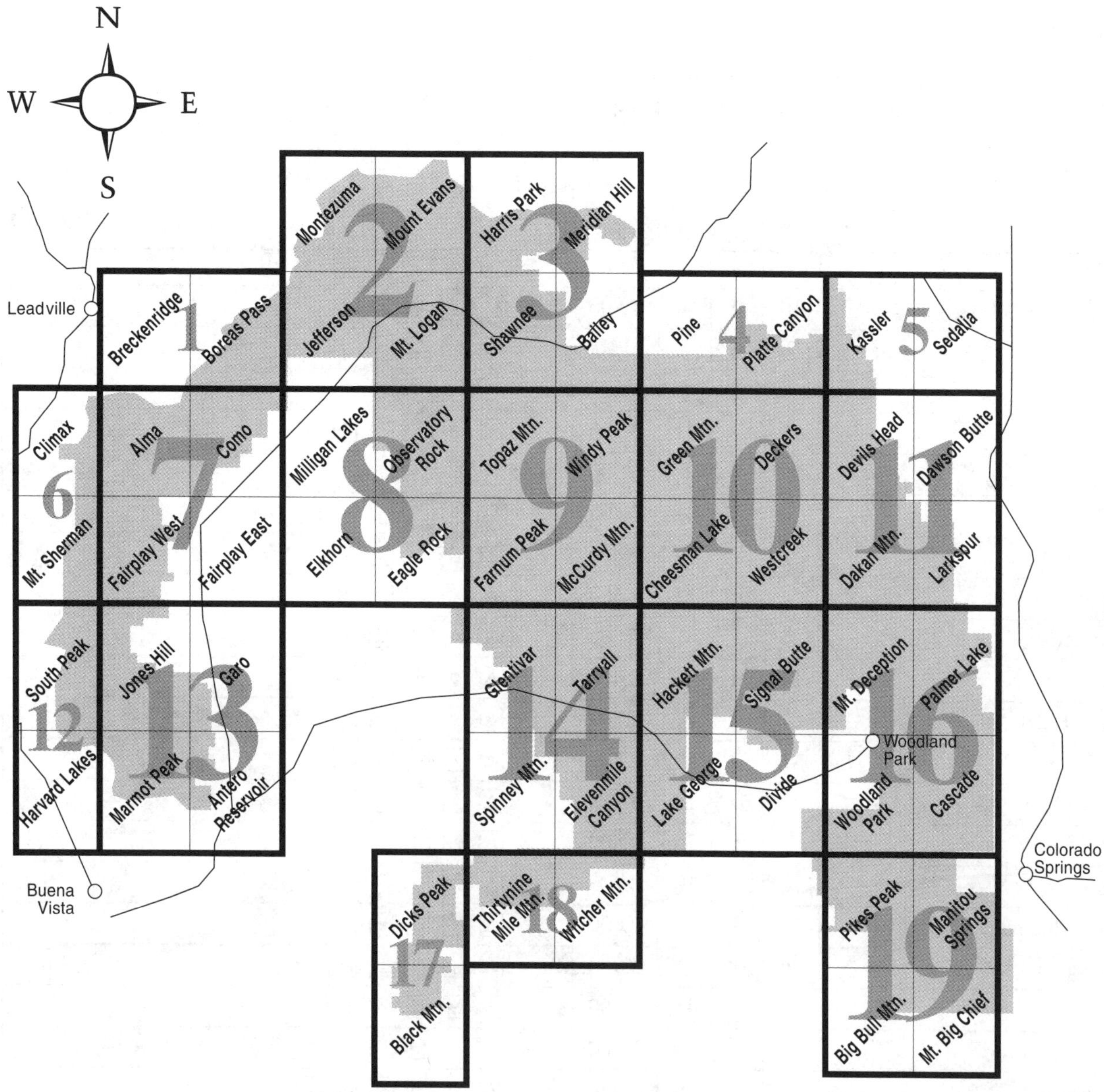

Map Numbers

Pike National Forest

(Showing USGS 7 1/2' quadrangle topographic map coverage)

Pike & San Isabel National Forest
Pikes Peak Ranger District
601 South Weber Street
Colorado Springs, Colorado 80903
(719) 636-1602

CAMPGROUND SITE INDEX

NAME OF SITE	MAP NUMBER	SITE NUMBER	NO. OF UNITS	ELEVATION	FACILITIES
Aspen	2	4	12	9,900	Campgrounds, Fee Required, Restrooms (barrier free), Drinking Water, Fishing, Hiking Trail, Bicycle Trail
Baldy	10	5	8	7,800	Campgrounds (barrier free), Restrooms (barrier free), Drinking Water, Bicycle Trail
Beaver Creek	7	2	3	10,900	Campgrounds, Restrooms, Fishing, Bicycle Trail
Big Turkey	15	5	10	8,000	Campgrounds, Restrooms, Drinking Water
Blue Mountain	15	2	21	8,200	Campgrounds, Fee Required, Restrooms (barrier free), Drinking Water, Hiking Trail
Bridge Crossing	10	8	6	6,300	Campgrounds, Restrooms, Fishing
Buffalo	10	3	41	7,400	Campgrounds (barrier free), Fee Required, Restrooms (barrier free), Drinking Water, Hiking Trail, Bicycle Trail
Buffalo Springs	13	1	17	9,000	Campgrounds, Restrooms (barrier free), Drinking Water, Hiking Trail
Burning Bear	2	10	13	9,500	Campgrounds, Restrooms, Drinking Water, Hiking Trail, Horse Trail
Colorado	16	2	81	7,800	Campgrounds, Fee Required, Restrooms, Drinking Water, Fishing, Hiking Trail, Bicycle Trail
Cove	14	4	5	8,400	Campgrounds, Fee Required, Restrooms, Drinking Water, Fishing
Devils Head	11	2	21	8,800	Campgrounds, Fee Required, Restrooms, Drinking Water, Hiking Trail, Motorized Trail
Flat Rocks	11	1	19	8,200	Campgrounds, Fee Required, Restrooms, Drinking Water, Motorized Trail
Four Mile	7	3	14	10,800	Campgrounds, Fee Required, Restrooms (barrier free), Drinking Water, Fishing, Hiking Trail, Bicycle Trail
Geneva Park	2	6	26	9,800	Campgrounds, Restrooms, Drinking Water
Goose Creek	10	12	10	8,100	Campgrounds, Restrooms, Drinking Water
Green Mountain	10	1	6	7,600	Campgrounds, Restrooms, Drinking Water, Bicycle Trail
Hall Valley	2	1	9	9,900	Campgrounds, Restrooms, Drinking Water, Hiking Trail, Horse Trail
Handcart	2	2	11	9,800	Campgrounds, Restrooms, Drinking Water
Happy Meadows	15	3	7	7,900	Campgrounds, Fee Required, Restrooms, Drinking Water, Fishing
Horseshoe	7	4	19	10,600	Campgrounds, Fee Required, Restrooms, Drinking Water, Fishing, Hiking Trail, Bicycle Trail, Motorized Trail
Indian Creek	5	1	11	7,500	Campgrounds, Fee Required, Restrooms, Drinking Water, Motorized Trail
Jackson Creek	11	3	9	8,100	Campgrounds, Restrooms, Drinking Water
Jefferson Creek	2	3	17	10,100	Campgrounds, Fee Required, Restrooms (barrier free), Drinking Water, Fishing, Hiking Trail, Bicycle Trail
Kelsey	10	6	17	8,000	Campgrounds, Fee Required, Restrooms, Drinking Water
Kenosha Pass	2	8	25	10,000	Campgrounds, Fee Required, Restrooms, Drinking Water, Hiking Trail, Bicycle Trail, Horse Trail
Kite Lake	7	5	7	12,000	Campgrounds, Restrooms, Fishing, Hiking Trail
Lodgepole	2	5	35	9,900	Campgrounds, Fee Required, Restrooms (barrier free), Drinking Water, Fishing, Hiking Trail, Bicycle Trail
Lone Rock	10	9	19	6,400	Campgrounds (barrier free), Fee Required, Restrooms (barrier free), Fishing
Lost Park	9	2	10	10,000	Campgrounds, Restrooms (barrier free), Drinking Water, Fishing, Hiking Trail, Bicycle Trail
Meadow Ridge	16	7	19	9,200	Campgrounds (barrier free), Fee Required, Restrooms, Drinking Water, Fishing, Boating, Hiking Trail, Bicycle Trail

White reversed-out symbol indicates barrier free facilities available

Campgrounds	Restrooms	Fishing	Hiking Trail	Motorized Trail
Fee Required	Drinking Water	Boating	Bicycle Trail	Horse Trail

CAMPGROUND SITE INDEX

NAME OF SITE	MAP NUMBER	SITE NUMBER	NO. OF UNITS	ELEVATION	FACILITIES
Meadows (Group)	10	2	300	7,000	Campgrounds, Fee Required, Restrooms, Drinking Water, Hiking Trail, Bicycle Trail, Horse Trail
Meridian	3	2	18	9,000	Campgrounds, Restrooms, Drinking Water
Michigan Creek	1	2	13	10,000	Campgrounds, Fee Required, Restrooms (barrier free), Drinking Water, Fishing, Bicycle Trail
Molly Gulch	10	11	15	7,500	Campgrounds, Restrooms, Drinking Water
Painted Rocks	16	1	18	7,900	Campgrounds, Fee Required, Restrooms, Drinking Water, Fishing, Hiking Trail, Bicycle Trail
Pike Community (Group)	16	3	350	7,700	Campgrounds, Fee Required, Restrooms, Drinking Water
Platte River	10	7	10	6,300	Campgrounds, Restrooms, Drinking Water, Fishing, Motorized Trail
Redrocks (Group)	16	5	90	8,200	Campgrounds, Fee Required, Restrooms, Drinking Water, Hiking Trail
Riverside	15	1	19	8,000	Campgrounds, Fee Required, Restrooms (barrier free), Drinking Water, Fishing, Hiking Trail
Round Mountain	14	1	16	8,500	Campgrounds, Fee Required, Restrooms (barrier free), Drinking Water, Bicycle Trail
Selkirk	7	1	15	10,500	Campgrounds, Restrooms, Fishing, Bicycle Trail
South Meadows	16	4	64	8,000	Campgrounds, Fee Required, Restrooms, Drinking Water, Fishing, Hiking Trail, Bicycle Trail
Spillway	14	5	24	8,500	Campgrounds, Fee Required, Restrooms, Drinking Water, Fishing, Hiking Trail
Springdale (Group)	16	6	30	9,200	Campgrounds, Fee Required, Restrooms
Springer Gulch	14	2	15	8,300	Campgrounds, Fee Required, Restrooms (barrier free), Drinking Water, Fishing
Spruce Grove	9	1	28	8,600	Campgrounds, Fee Required, Restrooms, Drinking Water, Fishing, Hiking Trail
The Crags	19	1	17	10,100	Campgrounds, Restrooms, Drinking Water, Hiking Trail
Thunder Ridge	16	8	21	9,200	Campgrounds (barrier free), Fee Required, Restrooms, Drinking Water, Fishing, Boating, Hiking Trail, Bicycle Trail
Timberline (Group)	2	7	200	9,800	Campgrounds, Fee Required, Restrooms
Top of the World	4	1	7	7 ,500	Campgrounds, Restrooms, Hiking Trail, Bicycle Trail, Motorized Trail
Trail Creek	15	6	7	7,800	Campgrounds, Restrooms, Hiking Trail, Bicycle Trail, Motorized Trail
Tramway	10	4	6	7,200	Campgrounds, Restrooms, Bicycle Trail
Wagon Tongue	14	3	7	8,400	Campgrounds, Restrooms, Fishing
Weston Pass	12	1	14	10,200	Campgrounds, Restrooms, Drinking Water, Fishing, Hiking Trail
Whiteside	2	9	7	8,900	Campgrounds, Restrooms, Drinking Water, Fishing, Hiking Trail, Horse Trail
Wigwam	10	10	10	6,600	Campgrounds, Restrooms (barrier free), Drinking Water, Hiking Trail
Wildhorn	15	4	9	9,100	Campgrounds, Restrooms, Drinking Water, Hiking Trail, Motorized Trail, Horse Trail

White reversed-out symbol indicates barrier free facilities available

Campgrounds	Restrooms	Fishing	Hiking Trail	Motorized Trail
Fee Required	Drinking Water	Boating	Bicycle Trail	Horse Trail

SECTION III

MTN BIKING

A. Rampart Reservoir

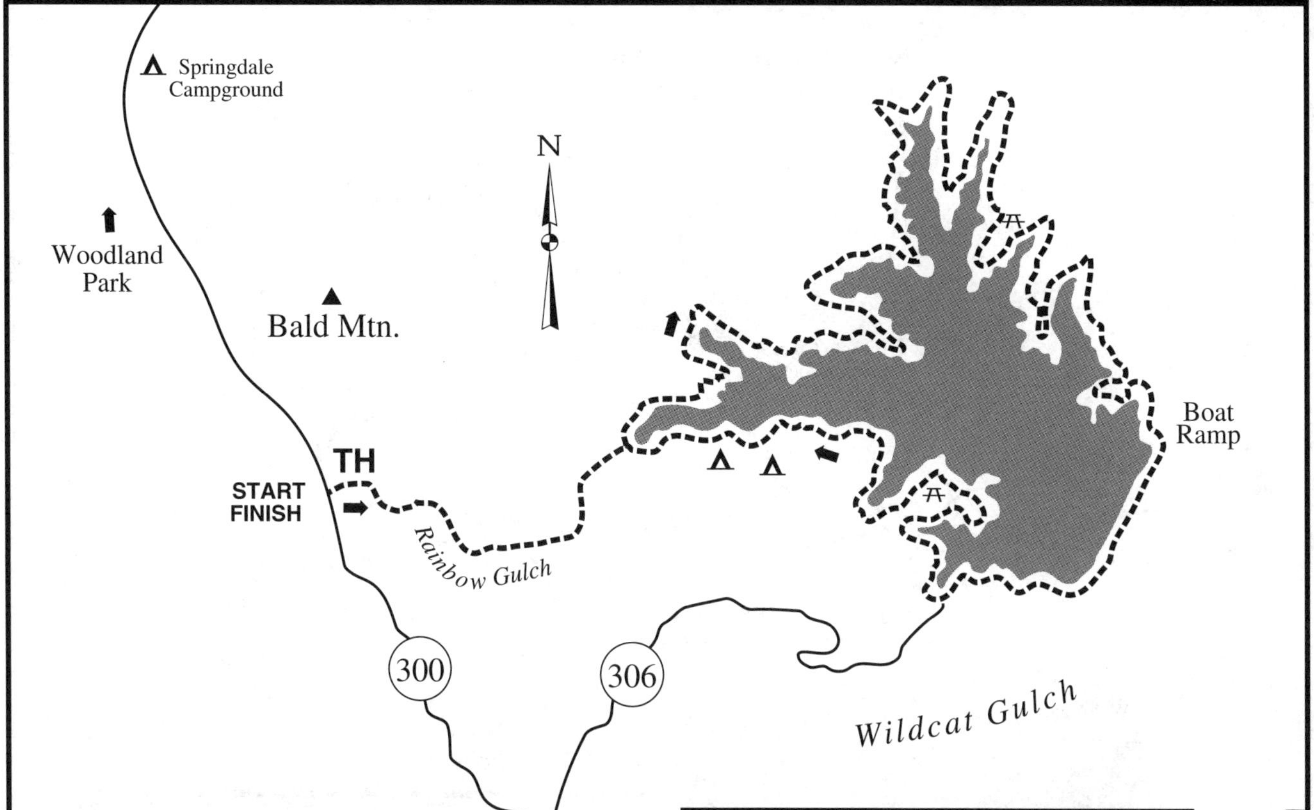

Distance:	**14.6 mile loop**
Time:	**2 to 4 hours**
Elevation Gain:	**380 feet (TH to reservoir)**
Difficulty:	**Moderate**
Surface:	**Trail**

■ Directions: From Woodland Park take County Road 23 (Rampart Range Road) for 2.9 miles to Loy Gulch road, right on Loy Gulch Road for 1.5 miles, right on Rampart Range Road 2.2 miles the trail head is on the left side of the road (approx. 6.6 mi.).

■ Ride: From the trailhead gate it's downhill to the pipeline spillway, go left here until you reach the bridge at 1.5 miles. An 11.6 ride around the reservoir with several small stream crossings, and a few short sections that are technically tricky. Turn left on the Rainbow Gulch Trail and return to the trailhead

■ Note: Picnic facilities, restrooms and camping, but there is no potable water.

Maps: USGS 71/2' Cascade and Woodland Park Quads
Information: Pike National Forest - Pikes Peak Ranger District (719) 636-1602

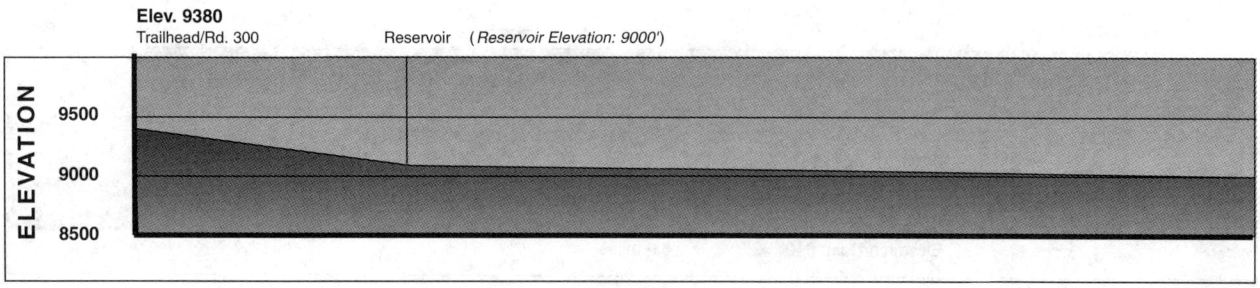

B. Buffalo Crk./Colo. Trail *Colo. Spgs/Denver*

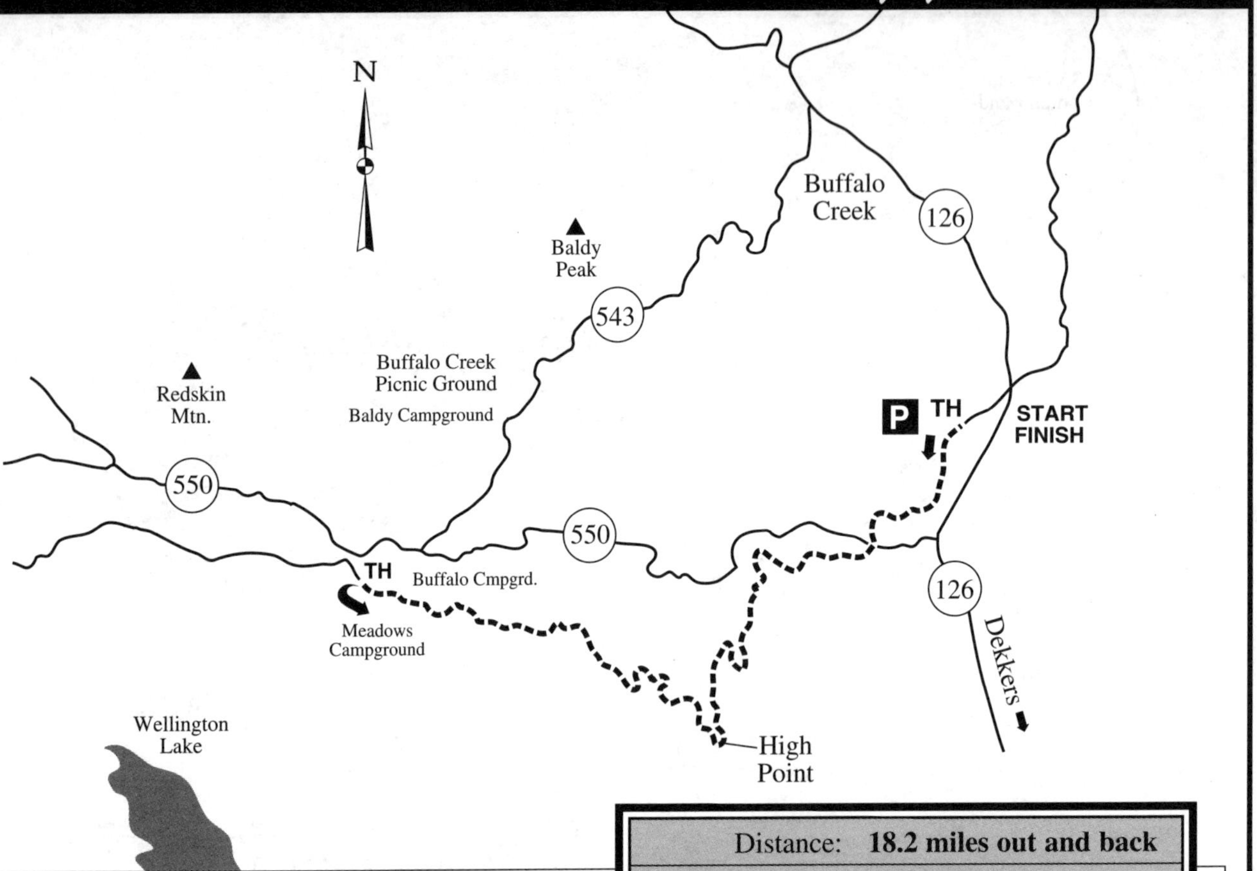

Distance:	**18.2 miles out and back**
Time:	**3 to 5 hours**
Elevation Gain:	**Approx. 2100 feet**
Difficulty:	**Moderate, but strenu-**
ous	
	exertion level
Surface:	**Primarily smooth**

■ Directions: From Denver drive south on US 285 to Pine Junction, turn left on Jefferson County Road 126 through Pine and Buffalo Creek. Continue for 3.5 miles to the Colorado Trail (1776) parking lot on the right.

■ Ride: Go west from the parking lot for 1/4 mile, turn left onto an old road that runs south a little over a mile. This is a good warm up for the singletrack that takes off to the right, winds around a rock outcrop, and crosses FS Road 550. The trail continues as a classic singletrack, descending a little before intersecting Shinglemill Trail. Go left and ride the rolls and rises, pass under large rock outcroppings (Litlle Scraggy Peal), cross a number of creek drainages before a gradual climb leads to where the trail joins an old logging road, which ends in a small aspen meadow. At the end of the meadow is Tramway Creek Trail. Go left, continuing on the Colorado Trail, diving down and across another (usually dry) creekbed, further to a narrow wooden bridge crossing and climb again to an open area. A newer section of singletrack rolls along and crosses an old road before the tightly curving descent to the Meadows Group Campground.

■ Information: (303) 636-1602

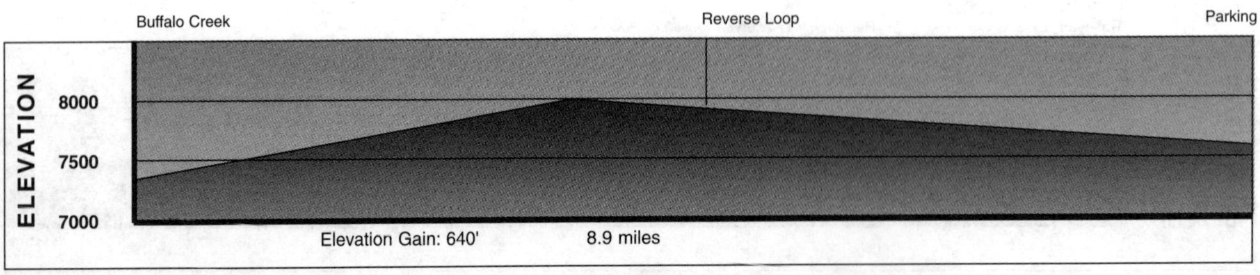

C. Morrison/Tram. Crk. Loop *Colo. Spgs/Denver*

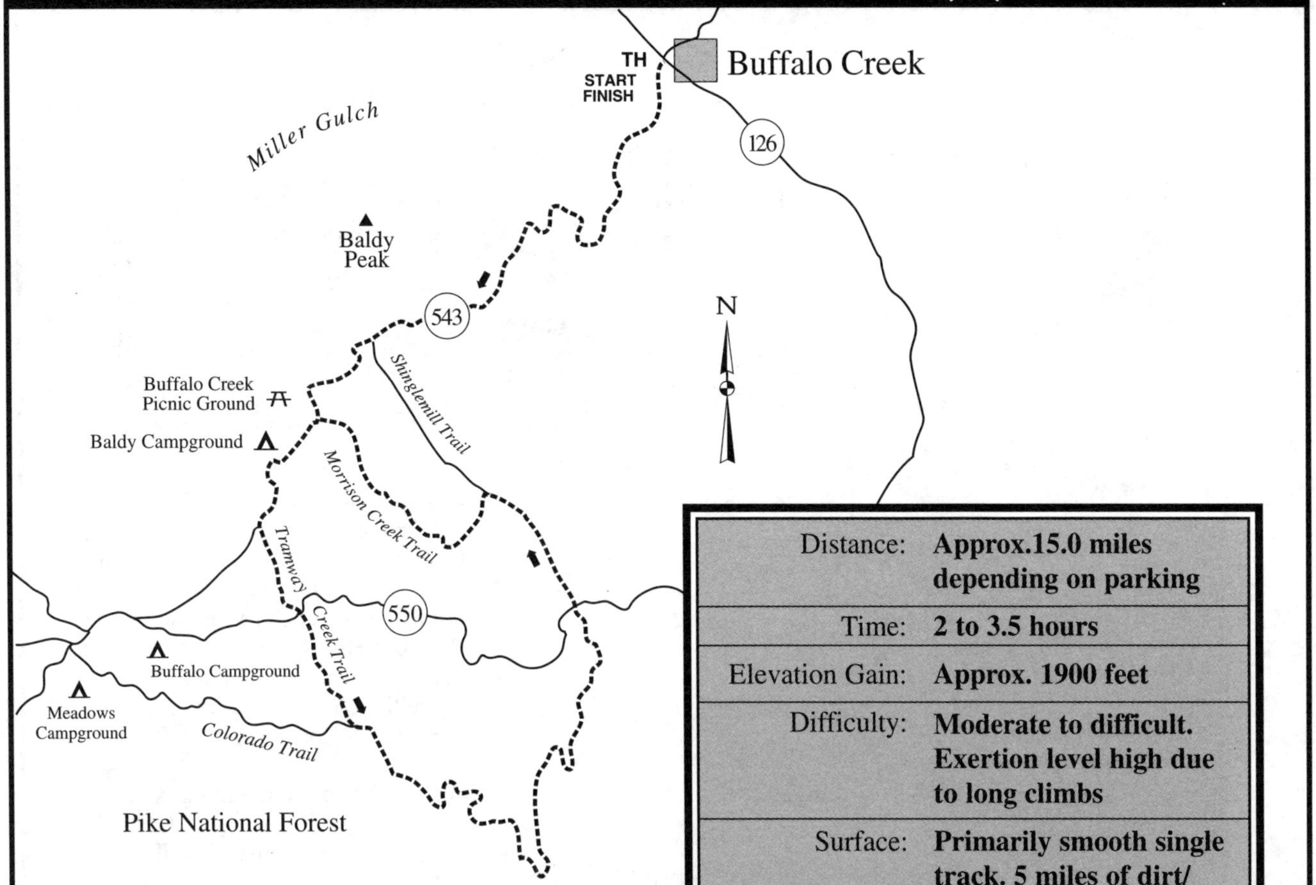

Distance:	**Approx.15.0 miles depending on parking**
Time:	**2 to 3.5 hours**
Elevation Gain:	**Approx. 1900 feet**
Difficulty:	**Moderate to difficult. Exertion level high due to long climbs**
Surface:	**Primarily smooth single track. 5 miles of dirt/ gravel road**

■ Directions: From Denver drive south on US 285 to Pine Junction, left (south) on Jefferson County Rd 126 through Pine and Buffalo Creek. Turn right on Wellington Lake Road (FS Road 543) about 1/2 mile from Buffalo Creek. Drive up FS Road 543 about 1/2 mile and park in a turn out.

■ Ride: Go west on FS Road 543 about 2.5 miles, passing Buffalo Creek Picnic Ground and Baldy Campground. Turn left onto the Morrison Creek Trail and begin a long steady climb up the Morrison Drainage for 2 miles to Shinglemill Trail. Go right on Shinglemill which is a milder ascent on new singletrack that dips and rolls until its termination at the Colorado Trail. Go right on the Colorado Trail, and ride along the single-track for three miles. After a long, gradual climb, the trail joins an old service road for a fast mile long descent (watch it! There are some sandy washes, exposed roots, and a sharp pitch through a rocky creek bed) in the Tramway Drainage. After riding through a meadow, go straight at the Tramway Creek Trail. The next 1.1 mile descent has a number of stream crossings. Tramway Creek crosses FS Road 550, and goes down the streambed, coming onto FS Road 543 at Tramway Creek Campground. To get back to start, get on the road on the right.

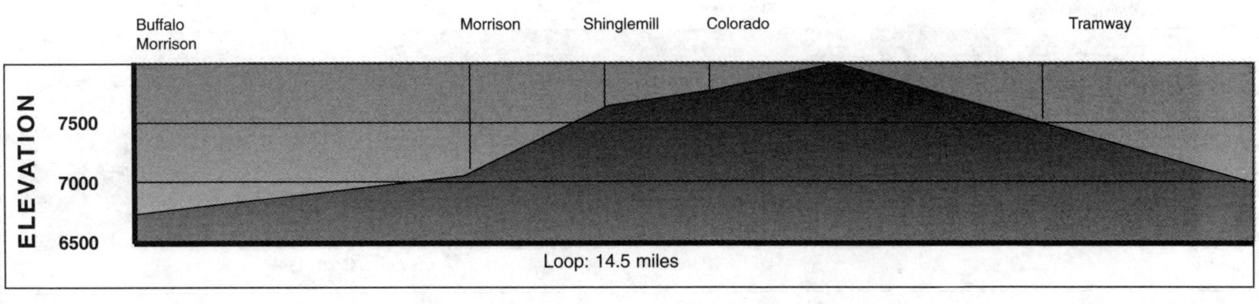

D. Stoney Pass

Colo. Spgs/Denver

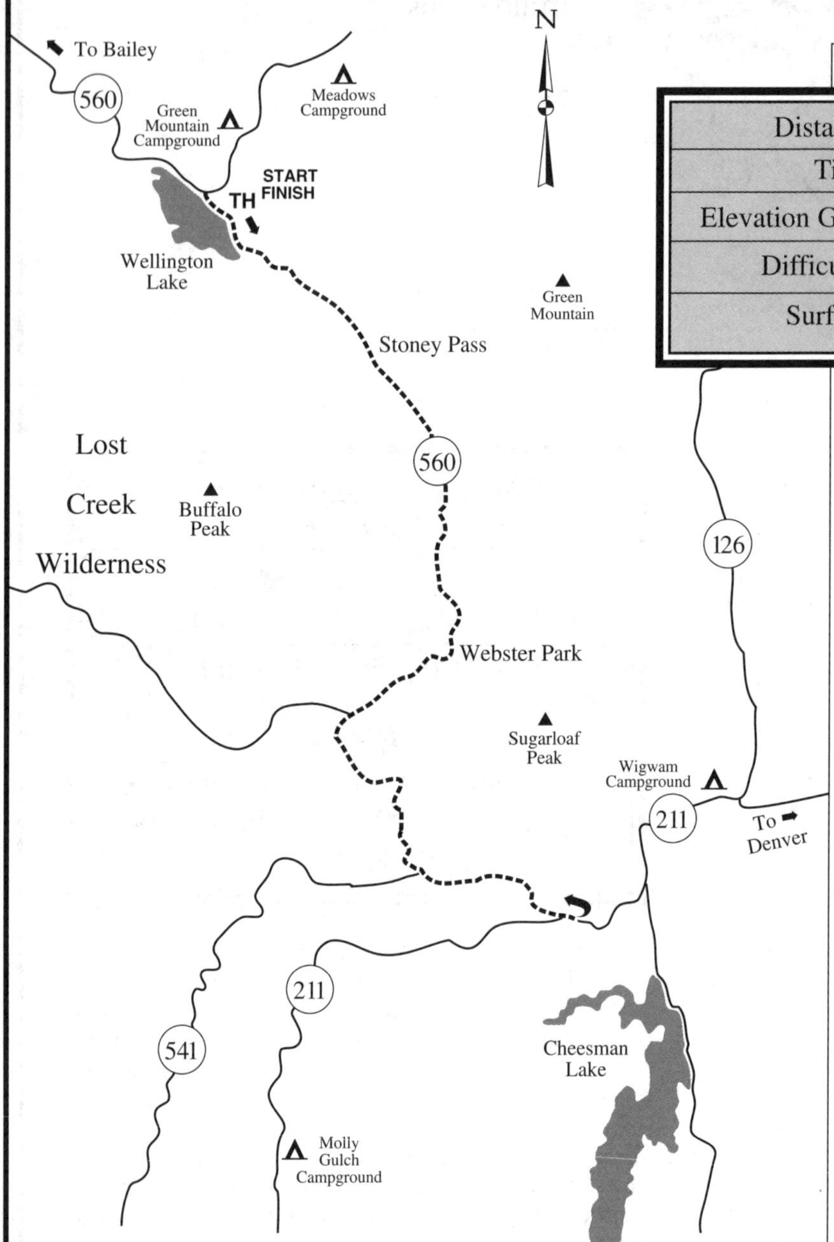

Distance:	**20.0 miles round-trip**
Time:	**3 to 4 hours**
Elevation Gain:	**1150 feet**
Difficulty:	**Moderate**
Surface:	**Paved road**

■ Directions: West from Denver on US 285 40 miles to Bailey, left on County Road 68, the road becomes FS Road 560. FS Road 560 is the first left as you enter Bailey at the bottom of Crow Hill. Continue on FS road 560 for 10 miles following the signs to Wellington Lake. Park at the cross-roads of FS Roads 543 & 560.

■ Ride: From Wellington Lake south on FS Road 560 along side of the lake. The road climbs steadily for 2 miles to the summit of Stoney Pass from the top of the pass it is a long gradual descent to Webster Park. A easy ride through the meadow, a short climb to the crest of a ridge, down a short steep grade to Wigwam Creek to FS Road 211

■ Note: A scenic lightly used route.

Maps: Pike National Forest map
Information: South Platte Ranger District (303) 236-5371

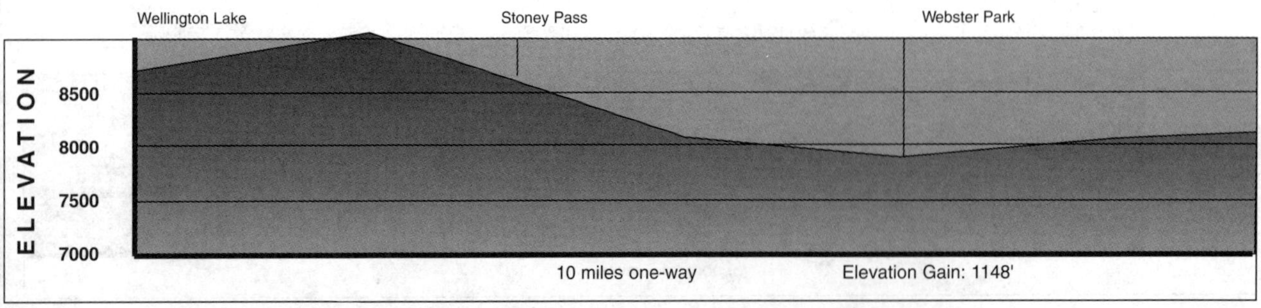

E. Waterton Canyon

Colo. Spgs/Denver

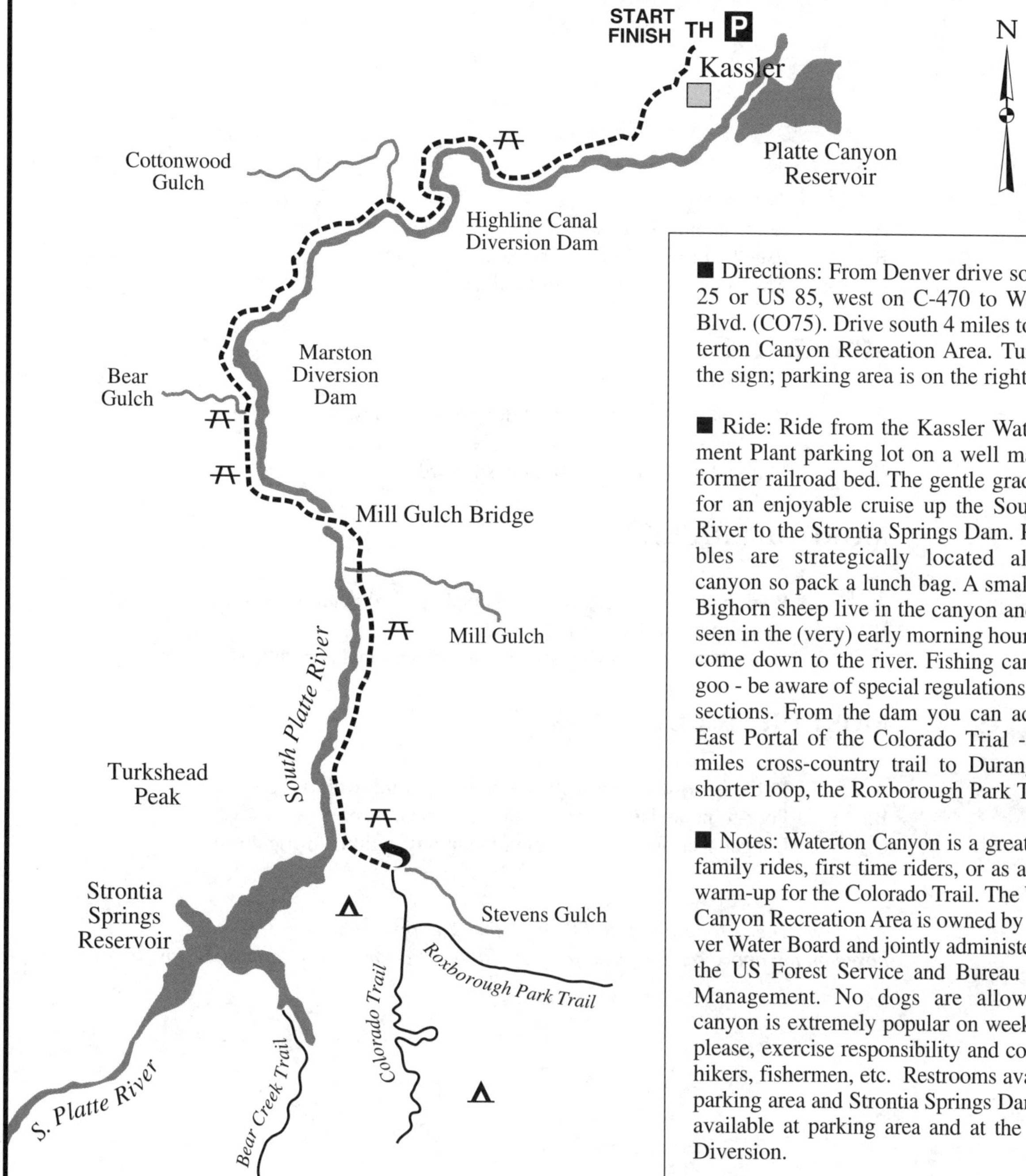

■ Directions: From Denver drive south on I-25 or US 85, west on C-470 to Wadsworth Blvd. (CO75). Drive south 4 miles to the Waterton Canyon Recreation Area. Turn left at the sign; parking area is on the right.

■ Ride: Ride from the Kassler Water Treatment Plant parking lot on a well maintained former railroad bed. The gentle grade makes for an enjoyable cruise up the South Platte River to the Strontia Springs Dam. Picnic tables are strategically located along the canyon so pack a lunch bag. A small herd of Bighorn sheep live in the canyon and best be seen in the (very) early morning hours as they come down to the river. Fishing can also be goo - be aware of special regulations on some sections. From the dam you can access the East Portal of the Colorado Trial - the 456 miles cross-country trail to Durango, or a shorter loop, the Roxborough Park Trail.

■ Notes: Waterton Canyon is a great area for family rides, first time riders, or as a six mile warm-up for the Colorado Trail. The Waterton Canyon Recreation Area is owned by the Denver Water Board and jointly administered with the US Forest Service and Bureau of Land Management. No dogs are allowed. The canyon is extremely popular on weekends, so please, exercise responsibility and courtesy to hikers, fishermen, etc. Restrooms available at parking area and Strontia Springs Dam. Water available at parking area and at the Marston Diversion.

■ Information: (303) 836-2031

Distance:	**12.5 miles out and back**
Time:	**1.5 to 3 hours**
Elevation Gain:	**1600 feet**
Difficulty:	**Easy**
Surface:	**Maintained gravel road**

METHOD FOR RATING TRAIL DIFFICULTY

Four categories for degree of difficulty are as follows:

EASY:

A. Route is most level with short uphill/downhill sections.
B. Excellent to good tread surface and clearance.
C. Absence of navigational difficulties/hazards.

MODERATE:

A. Route is level to sloping with longer uphill/downhill sections.
B. Good-to-fair surface and clearance.
C. Minimal navigational difficulties/hazards.

MORE DIFFICULT:

A. Route is level to steep with sustained uphill/downhill sections.
B. Fair to poor surface and clearance.
C. Short sections involving significant navigational difficulties/hazards.

MOST DIFFICULT:

A. Route is mostly steep with sustained uphill/downhill sections.
B. Poor-to-nonexistent tread surface and clearance.
C. Longer sections involving significant navigational difficulties/hazards.

Any rating (i.e., Moderate trail difficulty) found in this guide is based on the above scale which has been established for Forest Service purposes.

ALPHABETICAL TRAIL LISTING

Trails are in numerical order in document.

**** Trails open to trailbikes and ATV's less than 40" wide***

ORDERING
U.S. GEOLOGICAL SURVEY
TOPOGRAPHIC MAPS

For reference Outdoor Books & Maps has divided Pike National Forest into 19 maps, each map represents four 7 1/2' quadrangle topographic maps (Scale: 1" = 2000 Feet). These four color maps show in detail all physical features (Water drainage, roads, towns, etc.) and elevations. The elevations are graphically shown using contour lines spaced at 20 foot intervals. The index map in the front of the guide names the quads for each map, use the index map to locate the quads you wish to order.

Ordering topographic maps from the U.S. Geologic Survey:

The maps can be purchased for $2.50 per quad at the USGS map sales located in the Denver Federal Center at 6th and Kipling, Lakewood, Colorado. (Phone 1-800-435-7627).

Ordering topographic maps from Outdoor Books & Maps:

Outdoor Books & Maps accepts mail orders for topographic maps for all of Colorado's National Forests. Using the map index within the guide locate the area you wish to visit, select the topographic map by: Map number, quadrangle name, number of topographic maps times $2.50 for each map. Add shipping costs ($2.50) and sales tax if you are a Colorado resident.

A minimum order of four topographic maps ($10.00) is required - maps will be mailed folded to 8 1/2" X 11". Please allow 15 days for delivery.

ORDER FOREST SERVICE MAPS

Forest Service maps can be purchased at any Forest Service Office for $3.00 per map. At a small scale (1" = 2 miles) a four color map shows the entire forest. All the physical features are shown except elevations, the map gives an overview of the forest. For information on locations that sell these maps, call a Forest Service office listed in the front of this guide.

PIKE NATIONAL FOREST TOPOGRAPHIC MAP ORDER FORM

MAP #	QUADRANGLE NAME	QUANTITY	x $2.50	AMOUNT
	TOTAL			

Colorado residents add 7.2% sales tax: ____________

Shipping & Handling costs: **$2.50**

TOTAL AMOUNT: ____________

Please enclose payment to:

OUTDOOR BOOKS & MAPS
P.O. Box 417
Denver, CO 80201

Ship to: ____________

Address: ____________

City: ____________ *State:* ____ *ZIP:* ____